Christin Bechtel
Development of a MEMS-based
Confocal Laser Scanning Microscope
for Fluorescence Imaging

TUDpress

Christin Bechtel

Development of a MEMS-based Confocal Laser Scanning Microscope for Fluorescence Imaging

TUDpress
2015

Die vorliegende Arbeit wurde am 06. Januar 2014 an der Fakultät Elektrotechnik und Informationstechnik der Technischen Universität Dresden als Dissertation eingereicht und am 03. Juli 2014 verteidigt.

Vorsitzender der Promotionskommission:
Prof. Dr.-Ing. habil. Czarske

Gutachter:
Prof. Dr.-Ing. Lakner
Prof. Dr. rer. nat. Koch

Bibliografische Information der Deutschen Nationalbibliothek
Die Deutsche Nationalbibliothek verzeichnet diese Publikation in der Deutschen Nationalbibliografie; detaillierte bibliografische Daten sind im Internet über http://dnb.d-nb.de abrufbar.

Bibliographic information published by the Deutsche Nationalbibliothek
The Deutsche Nationalbibliothek lists this publication in the Deutsche Nationalbibliografie; detailed bibliographic data are available in the Internet at http://dnb.d-nb.de.

ISBN 978-3-95908-012-5

© 2015 TUDpress
Verlag der Wissenschaften GmbH
Bergstr. 70 | D-01069 Dresden
Tel.: 0351/47 96 97 20 | Fax: 0351/47 96 08 19
http://www.tudpress.de

Alle Rechte vorbehalten. All rights reserved.
Gesetzt vom Autor.
Printed in Germany.

The scientist does not study nature because it is useful, he studies it because he delights in it, and he delights in it because it is beautiful. If nature were not beautiful, it would not be worth knowing, and if nature were not worth knowing, life would not be worth living.

Jules Henri Poincaré

Acknowledgements

This thesis is the result of my PhD research at the Fraunhofer Institute for Photonic Microsystems (IPMS) and the Institute of Semiconductor and Microsystems at the Technische Universität Dresden. This research project was funded by the European Union and the Free State of Saxony (ESF) as "Landesinnovationspromotion" — I am grateful for this opportunity.

During this time, I was suppported by many people to whom I am very thankful.

Firstly, I would like to thank my doctoral thesis supervisor, Prof. Dr. Hubert Lakner, for his kind support and interest in this PhD topic, as well as for the opportunity to carry out my PhD research at the Fraunhofer Institute for Photonic Microsystems and the Institute of Semiconductor and Microsystems at the Technische Universität Dresden. He always encouraged me in my research. I would also like to thank Prof. Dr. Edmund Koch for accepting the co-supervision of this thesis.

In addition, I would like to thank Dr. Heinrich Grüger for his support and his continual attention to progress in the project. I would especially like to thank Dr. Jens Knobbe for sharing his in-depth understanding of optics with me, his guidance, advice and numerous scientific discussions which contributed significantly to this work.

I would also like to thank my colleagues at the Fraunhofer IPMS for the enjoyable time here. As well, I would like to thank my colleagues at the department of Optoelectronic Components and Systems, Technische Universität Dresden for warmly welcoming me and especially Hans-Jürgen Knoblauch for a lot of very helpful organizational support.

I would like to thank Prof. Dr. Günter Huber at the University of Hamburg for welcoming me and giving me the opportunity of carrying out measurements in his lab. In addition, I would like to thank especially Fabian Reichert for his organizational support and scientific support with lab equipment at the University of Hamburg. Additionally I would like to thank Prof. Dr. habil. Gunter Haroske from the Pathologic Institute 'Georg Schmorl' at the Krankenhaus Dresden-Friedrichstadt, Prof. Dr. Jörg Kriegsmann from the Group Practice for Pathology in Trier and Hans-Jörg Dethloff for providing me biological test samples.

Furthermore, my thanks to Prof. Dr. Eberhard Schultheiß for his support and having encouraged me to begin a PhD. To my partner and closest friends who were always there even in stressful times and supported me, my special thanks. Most importantly, I am very grateful to my parents, who always encouraged me and believed in me.

Funded by the European Union and the Free State of Saxony, Germany. - Finanziert aus Mitteln der Europischen Union und des Freistaates Sachsen

Zusammenfassung

Die konfokale Fluoreszenzmikroskopie erlaubt die selektive Darstellung einzelner Schichten von an sich "dicken"biologischer Proben und hat sich in der biologischen und medizinischen Forschung weltweit etabliert. Auch für den Einsatz im Bereich der klinischen dermatologischen Diagnostik ist die Verwendung der konfokalen Fluoreszenzmikroskopie hochinteressant, wie klinische Studien in den letzten Jahren gezeigt haben[1]. Mit diesem Mikroskopieverfahren werden tiefenaufgelöste Untersuchungen der oberen Hautschichten direkt am Patienten (in-vivo) ermöglicht. So lassen sich insbesondere das Penetrationsverhalten aufgetragener Substanzen visualisieren und der Verlauf dynamischer Hautprozesse untersuchen. Trotz des großen potenziellen Nutzens im Bereich der klinischen Diagnostik wird die konfokale Fluoreszenzmikroskopie dort bisher kaum angewendet. Wesentliche Faktoren die der Etablierung in der Dermatologie entgegen stehen, sind vor allem große Geräte, kleine Bildfelder und hohe Anschaffungskosten.

In der vorliegenden Doktorarbeit wird der neue Ansatz der Integration eines MEMS-Scannerspiegels zur Laserstrahlablenkung in einem konfokalen Fluoreszenzmikroskop untersucht und funktional überprüft. Zwei MEMS-basierte Fluoreszenz Laser-Scanning-Mikroskope wurden designt und werden vorgestellt, wobei das erste zum Funktionsnachweis praktisch aufgebaut und untersucht wurde.

Das erste Mikroskopsystem besteht vorwiegend aus kostengünstigen Kataloglinsen. Dieses System weist eine für diese Anwendung adäquate optische Performance auf. Die Möglichkeit tiefenaufgelöste Schnittbilder aufzunehmen wurde anhand von histologischen und biologischen Proben nachgewiesen. Eine quantitative Performanceanalyse und ein detaillierter Vergleich mit den Simulationsergebnissen wird durchgeführt. Begrenzungsfaktoren der Bildfeldgröße und der lateralen Auflösung werden diskutiert.

Die Transformation und die Optimierung des ersten Designs hin zu einem portablen, handlichen Mikroskop mit einem erweiterten Bildfeld wurde in einem zweiten Mikroskopsystem untersucht. Es wird gezeigt, dass dieses Ziel mit einem optimierten Optikdesign unter Verwendung von speziell angefertigten Linsen erreicht werden kann. Die Simulationsergebnissen zeigen, dass es möglich ist ein kompaktes und handliches Mikroskop mit einer um mehr als doppelt so großen Bildfeldfläche zu realisieren.

Diese Doktorarbeit zeigt, dass die Integration eines MEMS-Spiegels zur Laserstrahlablenkung in konfokalen Mikroskopen ein sehr vielversprechender Ansatz ist, der es erlaubt große Bildfelder und kleine Systemgrößen miteinander zu vereinen.

---

[1]Lademann et al., University Medical School in Berlin

Abstract

Although fluorescence confocal laser scanning microscopy is a widely used technique in biology, these microscopes are at present uncommon in medical diagnostics. However laser scanning fluorescence microscopy has shown to be a promising non-invasive imaging technique that allows depth resolved investigations of skin disorders in-vivo[2]. Compared to conventional wide field microscopy, this imaging technique has a clear advantage by allowing the evaluation of dynamic skin processes and penetration studies. Factors which have impeded the establishment of this technology so far are small fields of view and high costs.

The main objective of this dissertation has been to investigate and operationally verify a new approach, that of integrating a MEMS mirror for laser scanning. A compact fluorescence laser-scanning microscope with a field of view larger than that of currently available endoscopes is presented as an initial system layout. This first microscope design is mainly composed of off-the-shelf optics. The adequate optical performance and optical sectioning capability is illustrated with histological and biological samples. A quantitative optical performance evaluation and comparison with simulation results for this initial microscope is given. Limitations to the field of view and lateral resolution are discussed.

As secondary objective of this work the optimization and transformation of the initial design into a handheld version with enlarged field of view is investigated. It is shown that this goal can be achieved by re-designing of the optical imaging system including custom-designed lenses. Based on simulations, a handheld-version of the microscope with a more than twice as large field of view is possible (0.75 mm × 0.75 mm). In conclusion, this dissertation proves that the MEMS-based solution is a highly promising approach thanks to its ability to provide a large field of view while allowing a compact layout.

---

[2]Lademann et al., University Medical School in Berlin

Contents

1 Introduction   **1**
   1.1 Fluorescence Confocal Microscopy as an Imaging Tool 1
   1.2 Research Objective and Dissertation Overview 3

2 Background: Fluorescence Confocal Laser-Scanning Microscopy   **5**
   2.1 Fundamentals of Laser-Scanning Microscopy 5
   2.2 Promising F-LSM Applications in Dermatology 11
   2.3 Impact of Fluorescent Dyes on Microscopy 14
      2.3.1 Fluorescence Contrast Agents in Medicine 15
      2.3.2 Influence of Fluorescein on the Laser Illumination 16
   2.4 Introduction to the Wavefront Aberration Theory 20
   2.5 Important Figures of Merit for Image Quality Evaluations 24

3 MEMS Mirror Based F-LSM Design   **27**
   3.1 MEMS mirror based Implications on the Microscope Specifications 27
      3.1.1 Implications on the Illumination Optics 28
      3.1.2 Requirements on the Imaging Optics and the Z-Shifter 31
      3.1.3 Fluorescence Detection Path 34
   3.2 Intermediate Relay Optics - Performance Simulations with Zemax 39
   3.3 Analysis of Assembly Tolerances 43
   3.4 Dynamic Deformations of the MEMS Mirror 47
   3.5 Image Distortion - Causes and Correction Methods 49

4 F-LSM System Realization   **55**
   4.1 F-LSM Demonstration System Layout 55
   4.2 Quantitative Optical Performance Testing 64
   4.3 Wavefront Error Measurement with a Shack-Hartmann Sensor 72
   4.4 Images of Biological Samples 80
   4.5 Summary 84

5 Transfer into a Handheld Microscope Design   **87**
   5.1 Optimized Imaging Optics 87
      5.1.1 Scan Lens Design 89

	5.1.2 Tube Lens Design	94
5.2	Mounting and Assembly Tolerances	98
5.3	Summary and Comparison of Both F-LSM Systems	104

6 Discussion and Outlook   **107**

Appendix A: Flow Chart   **111**

Appendix B: Scan Lens - Technical Drawings   **113**

Abbreviations   **117**

List of Variables   **119**

List of Figures   **121**

List of Tables   **127**

Bibliography   **129**

1 Introduction

1.1 Fluorescence Confocal Microscopy as an Imaging Tool

Fluorescence confocal laser scanning microscopy (F-LSM) was invented by M. Minsky in 1957 and is now a widely used and powerful imaging technique in basic biological and medical research [1]. In contrast to optical wide-field microscopes, fluorescence confocal microscopes allow to perform depth-resolved fluorescence imaging of 'thick' specimens with increased imaging contrast by effectively suppressing out-of-focus light. This ability is often referred to as optical sectioning of 'thick' specimen. Post processing of the acquired serial cross-sectional images at defined depths provides 3D visualization of biological structures. The basic principle of laser scanning confocal microscopy is point-wise illumination of the sample with laser light and subsequent detection of the emitted fluorescence light. The confocal character of these microscopes consists of spatial filtering of the fluorescence signal collected from the specimen plane. To obtain 2D images, a scanning system is required to scan the focused spot across the sample. In general, this is achieved by directing a collimated laser beam onto one or two mirrors, which angularly deflect the beam in two directions. The subsequent telescope optics and objective, convert this angular movement of the scanning beam to lateral movement of the focal spot in the sample. Fluorescence light generated in the specimen is emitted in all directions. A fraction of the emitted fluorescence light is descanned along the original illumination path and directed onto a detector. This detector is placed behind a pinhole, that prevents out of focus light reaching it. Thus, spatial filtering of the signal collected from the specimen plane is ensured. For image reconstruction, the measured intensity and the position information of the mirror are correlated. In conclusion, the ability to resolve depth resides in the use of the same path for illumination as for detection, and in the spatial filtering of the fluorescent light.

Thanks to this advantageous capability of selectively examining cross-sections in a non-invasive manner at a defined depth parallel to the surface, this technique is a promising tool in dermatologic diagnostics. Time consuming and often inconvenient biopsies that are then followed by histological analysis of the tissue can be avoided with this technique. The only preparation needed for F-LSM imaging of the skin is to apply a fluorescent dye to stain the different cellular structures and membranes. This is necessary due to the fact

1 Introduction

that autofluorescence would be too low. An additional benefit of F-LSM imaging is that it not only allows for visualizing the different cell layers, it also enables physicians to perform penetration studies and to evaluate healing processes quantitatively. Furthermore, there is reasonable possibility that ongoing research in the field of fluoresceine-labeled antibodies for immunological diagnostics will expand the field of applications for F-LSM significantly in the clinical setting.

So far, there is no imaging technique available to medical practitioners for performing this microscopic inspection of the upper skin layer *in vivo*, nor do any available techniques enable these kind of penetration studies to be performed. The established depths resolving imaging techniques used in dermatology are reflectance confocal laser scanning microscopy (R-LSM) and optical coherence tomography (OCT). In reflectance confocal laser scanning microscopy, the scattering properties of the tissue is used for imaging. Contrast is provided by the different reflectivities of different structures in the tissue, e.g. organelles and cytoplasmatic melanin. However, at normal incidence, approximately four to seven percent of the incident light at all wavelengths reflects at the tissue surface because of the large change in the refractive index between the air and the tissue [2]. Thus, R-LSM, in contrast to F-LSM is not able to image the skin surface itself. With OCT on the other hand, greater imaging depth can be achieved than with F-LSM, but the resolution is not high enough for sub-cellular information. Furthermore, neither of these two techniques is able to image the penetration of substances into or within the tissue. As a result, F-LSM provides considerable benefits. However, despite these several benefits of F-LSMs as non-invasive diagnostic tool, its use so far is uncommon in clinical practice. This is mainly due to the fact that the fluorescence confocal laser scanning microscopes used in medical and biological research are generally complex, stationary bench-top systems, that are therefore costly (>150,000 EUR). These factors impede their establishment in medical practice, where affordability and portability are of greater importance. In addition, the common F-LSM's used in research up to now are based on a pair of galvanometer mirrors for laser beam scanning. Even though these F-LSM's provide high-speed scanning, a reduction of their size is limited by the relatively large galvanometer mirror modules. For this reason, several approaches have been pursued to develop a compact F-LSM within the last ten years. Among them, fiber optic scanning endoscopes are prominent examples [[3],[4],[5],[6],[7]]. These endoscopes are designed for *in-vivo* imaging inside the human body and possess a restricted field of view as a result. An alternative approach is the integration of a micro-electro-mechanical system (MEMS) mirror as a scan unit [[8],[9],[10],[11],[12],[13],[14]]. MEMS mirrors are small, lightweight and mechanically stable while possessing high reflectivity. In addition, their scan angles are larger than those of fiber optic scanning techniques and therefore allow for a wider field of view. Nevertheless, the reported systems based on MEMS mirrors only possess moderate fields of view. This is mainly due to their field of application, the mirror choice and the imaging optics. However, for dermatological applications, a large field of view is of great interest. It is for this reason that a dual-axis MEMS mirror provided by the Fraunhofer Institute of Photonic Microsystems is used for this PhD thesis.

1.2 Research Objective and Dissertation Overview

The main objective of the research here presented is to design and develop a compact fluorescence confocal laser-scanning microscope (F-LSM) with a large field of view that possesses adequate resolution for clinical dermatologic diagnostics. For this design, (1) a compact scanning mechanism, (2) highly corrected but affordable optical system for fluorescence imaging, as well as (3) a space-saving layout and compact packaging are necessary. As secondary objective of this work the optimization and transformation of the initial design into a handheld version with an enlarged field of view is investigated. The target specifications for the microscope system derive from dermatologic diagnostics. In the second chapter of this PhD thesis, the background for this research, including fundamentals of confocal imaging, a brief review of state-of-the-art scanning techniques in confocal imaging systems, an introduction to the basics of fluorescence, the use of fluorescence dyes in medicine, and their impact on the F-LSM design will be outlined. A comparison of the MEMS mirror utilized and state-of-the-art scan techniques will be given. The optical design of a small bench-top fluorescence confocal microscope demonstration system based on a MEMS mirror with corresponding figures of merit, optical simulation results and a corresponding tolerance analysis is presented in chapter 3. Here, implications and requirements resulting from the integration of the MEMS mirror on the remaining optical components are discussed and correlated figures of merit for the microscope performance are outlined. This bench-top system is used to verify that it is feasible to build up a MEMS based large field of view F-LSM mainly composed of off-the-shelf components with an appropriate subcellular resolution for fluorescence confocal imaging in dermatology. In chapter 4, the construction and characterization of this demonstration set-up are described and a comparison to the predicted performance analysis in chapter 3 is given. For quantitative analysis of the optical performance resolution targets and wavefront measurements are used and the results are compared with prior simulations. Images of biological samples and histological probes are presented. Chapter 5 presents a further miniaturization of the demonstration set-up with the integration of a customized lens design. Performance simulations of the optical design, the impact of geometrical constraints, as well as mounting tolerances are analyzed and final assembly and manufacturing specifications for the lens and the overall system are presented. This final version of the dual-axis MEMS-based F-LSM would meet the target specifications of resolution, space requirements and affordability for application in dermatology. In chapter 6 the limitations of this F-LSM system and the integrated dual-axis MEMS

1 Introduction

mirror are discussed and future path for further miniaturization and improvement to the performance of the system are given. In conclusion, prospects for further research and a summary of the whole thesis are given.

2 Background: Fluorescence Confocal Laser-Scanning Microscopy

In the first section of this chapter, an introduction to confocal laser-scanning microscopy is given and state-of-the-art scanning techniques will be reviewed. In the next section, the potential for and application of fluorescence laser-scanning microscopy in dermatological diagnostics will be introduced and an outline of recent research activities will be given. For this application, it is necessary to apply a fluorescent dye prior to the diagnostic procedure. Appropriate fluorescent dyes approved for use in humans, their characteristics, and resulting demands placed on the confocal microscope illumination will be introduced and discussed. In the last section of this chapter an introduction to the wavefront aberration theory and important figures of merits for optical performance evaluations are introduced.

2.1 Fundamentals of Laser-Scanning Microscopy

Confocal microscopy with a point source for illumination was invented by M. Minsky in 1957 [1]. The most basic form is depicted in figure 2.1a. In contrast to wide-field microscopy, confocal microscopy is a sequential imaging technique in which only information from the focal region is detected and the image is built up point by point. What is termed as the confocal character of the microscope follows from the placement of a pinhole in front of the detector in a plane which is conjugate to the illumination pinhole or point source of light. Out-of-focus light is prevented by this arrangement from reaching the detector, as depicted in figure 2.1. The resulting depth-resolved imaging is often referred to as optical sectioning of a 'thick' specimen. To get a full 2D image with this sequential imaging technique, it is necessary to move the focal spot, or focus, within the specimen. In the early confocal microscopes, this was achieved by moving the mounted specimen in the focal plane, which is referred as stage scanning. Although a flying-spot microscope was already patented in 1952 by Young and Roberts [15] and confocal microscopes with laser illumination in the seventies by Egger and Davidovits [[16],[17]] laser scanning was not applied in confocal microscopes until the eighties. Finally, the integration of both laser illumination and laser scanning in a confocal microscope was realized by different groups nearly in parallel [[18],[19],[20]]. In the first system, White

2 Background: Fluorescence Confocal Laser-Scanning Microscopy

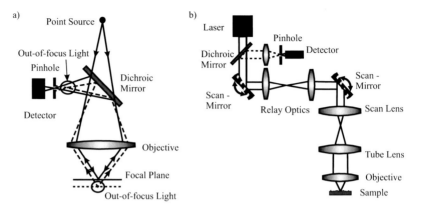

Fig. 2.1: Optical path of a) a confocal microscope with stage scanning per Minsky, 1957, b) confocal laser-scanning microscope using two scanning mirrors

et al. used a rotating polygonal mirror for laser scanning, but soon replaced it by a pair of oscillating galvanometric mirrors, a technique which is still one the most common applied in confocal microscopes today [21]. As depicted in figure 2.1b, in this type of confocal laser-scanning microscope, a collimated laser beam is directed onto the mirrors that deflect the beam in two directions. When using two mirrors, they are either placed in close proximity to each other or one is imaged onto the second by relay optics. With the subsequent telescope optics and the objective adjacent to the mirrors, the angular movement of the laser beam is transformed into lateral movement of the focal spot in the sample. During detection, the backscattered light is collected by the objective and returns along the original illumination path, which is referred as descanning [[22],[23]]. Due to the fact that the speed of the scanning mirrors is very slow compared to the speed of light, the optical descanning path is identical to the original path of the illumination beam. Finally, out-of-focus light is prevented from reaching the detector by a pinhole placed in front of it.

The development of laser illumination and laser scanning was accompanied by new areas of application. Traditional reflective imaging was supplemented by the application of confocal scanning microscopes to fluorescence imaging in biological research. Instead of detecting a portion of backscattered light from the surface, fluorescence laser-scanning microscopy uses laser illumination to generate fluorescence in the sample that is then detected. Excepting for the necessity of color-correcting (achromatic) optics and filters to separate the illumination and fluorescent light, the basic principle of these microscopes has hardly changed from the principle initially described by Minsky. The microscope set-up still consists of four functional groups: the illumination unit, the scanning unit,

2.1 Fundamentals of Laser-Scanning Microscopy

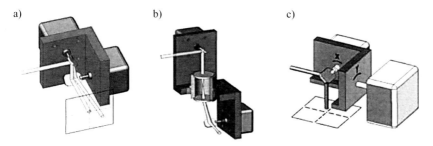

Fig. 2.2: Configurations of confocal microscope scanning system using galvanometric scanning mirrors [24]

the focusing optics and the detection unit, to provide point-wise illumination of the sample and subsequent confocal detection of the reflected and emitted fluorescent light from the focal area, with a pinhole in front of the detector.

Since then, however, great improvements have been achieved in the applied optics and much effort has been put into developing smaller, faster, and more-accurate galvanometric scanning mirrors. Besides galvanometric mirrors, alternative scan techniques such as MEMS mirrors and fiber-based scan units have been presented in the last ten years. However, the majority of commercially available confocal laser-scanning microscopes are still based on using galvanometric mirrors for laser scanning. In general, these microscopes are highly complex and consequently costly benchtop units (> 150,000 EUR) whose applications range from biomedical research to industrial material testing. The set of galvanometric mirrors included here are usually driven either linearly, resonantly, or a combination of both. For linearly driven mirrors, the data acquisition rate is limited to one pixel per two to three microseconds due to inertia [22]. Resonant galvanometers are used to achieve higher scanning speeds, usually operating at 4 kHz and higher [25]. However, the increase of scanning frequency is accompanied by a decrease of the pixel dwell time and thus a decrease in the amount of fluorescence signal falling on the detector. As a result, these higher frequencies cannot be applied for all types of measurements. Different configurations of the galvanometer mirrors are also frequently used. In general, the two mirrors are positioned orthogonal to each other. Their axes of rotation are perpendicular to each other so that one scanner moves the laser beam in the lateral direction, while the other one moves it in the vertical direction. As depicted in figure 2.2 a) and b), the mirrors can either be arranged in a close-coupled configuration or in a configuration where the mirrors are separated by relay optics. The latter configuration is the most common arrangement implemented in confocal microscopes thus far [26]. Here, one mirror is imaged by relay optics onto the second mirror, such that both mirrors are located in conjugate planes to the rear focal plane of the objective [[27],[26]]. For pur-

2 Background: Fluorescence Confocal Laser-Scanning Microscopy

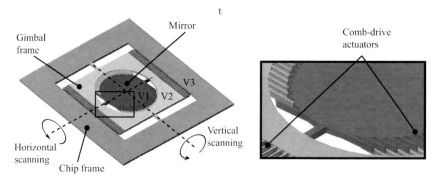

Fig. 2.3: Gimbal-mounted dual-axis MEMS mirror with comb-drive actuators [28]

poses of simplicity and to achieve a more compact size, the additional relay optics have been removed in recent microscope systems. In the close-coupled configuration depicted in figure 2.2 c), the scanning mirrors are placed as closely as possible to each other, with their geometric midpoint located in a conjugate plane to the rear focal plane of the objective. Thus, these two mirrors are only approximately positioned in the conjugate plane. This leads to slight movement of the laser beam at the entrance aperture of the objective and results in non-uniform illumination of the specimen, especially for large scan angles. To overcome this disadvantage, cardanic scanning systems were presented, whereby a small galvanometer head is mounted on a second larger galvanometric scanning device whose rotational axis passes through the mirror center perpendicular to the mirror axis as shown in figure 2.2 c). Despite the improvements in scanning frequency and mirrors' alignments, the size of the scan unit using either one or two galvanometric mirrors is still too large for handheld solutions. To transfer fluorescence confocal laser-scanning microscopy into a clinical setting, a cost-effective solution with a high degree of miniaturization is necessary. To achieve the necessary miniaturization, several approaches have been pursued within the last 10 years. Among these are confocal microscope systems based on dual-axis MEMS mirrors, which are currently still subjects of research [[8],[9],[10],[11],[12],[13],[14]].

Compared to galvanometric mirrors, MEMS mirrors are small in size while offering scan angles up to $\pm 15°$. Furthermore, they allow scanning both axes with a single mirror. An example of such a gimbal-mount MEMS mirror is the mirror from Fraunhofer IPMS used in this project, as depicted in figure 2.3. With gimbal mounting, both axes of the mirror plate can be scanned simultaneously. This is referred to as cardanic scanning. In contrast to galvanometric mirrors, MEMS mirrors are formed from single crystal of silicon with a bulk micromachining process. A comb drive is implemented to electrostatically drive the MEMS mirror [29]. The electrostatic force of the comb drive originates from capacitance

2.1 Fundamentals of Laser-Scanning Microscopy

changes between the comb fingers of the frame and those of the mirror plate, which are both located in the plane of the mirror. A square wave voltage is applied as drive signal at the maximum mechanical deflection angle to generate an electrostatic force between the driving electrodes. This leads to an acceleration of the plate back towards its rest position. The voltage is turned off the mirror passes its rest position and is turned on again at the next maximum angle of deflection. For simultaneous two-dimensional deflection, both axes are independently excited, leading to a sinusoidal motion of the mirror along both axes. This results in a Lissajous scan pattern, where the filling factor is determined by the driving frequencies selected. Adjusting the drive voltage and frequency allows the oscillation amplitudes and phase differences to be controlled. Typical frequencies for a 2D MEMS mirror with a mirror plate diameter of 2 mm and a scanning angle of 7.5°, are 199 Hz and 1.337 Hz for the slow and fast axes, respectively. Since the oscillation of the mirror is extremely stable, there is depending on the required resolution, no immediate need to measure the mirror orientation in closed-loop operation [30]. However, to increase the repetitive precision of the image, real-time information feedback about the mirror position would need to be implemented. As a result, it is possible to achieve high scanning resolution with MEMS mirrors when an appropriate feedback system is used. Reliability was tested during long-runs under controlled environmental conditions with the result that no changes to the resonant frequency or the oscillation amplitude were observed [30]. Besides these promising scanning characteristics, MEMS mirrors have been proven to be highly robust and shock resistant [30]. This is mainly due to the fact that the mirrors are formed from a single crystal of material, which leads to nearly ideal elastic behavior and high mechanical fracture strength. Hence, besides their main advantage of a minuscule outline, MEMS mirrors also meet the requirements for a reliable and robust set-up. In consequence these MEMS mirrors are a promising scanning device for miniaturized and portable scanning systems where large scanning angles are necessary. In addition, an aluminum coating can be applied to provide high reflectivity ($\approx 90\%$ at 633 nm[31]).

Besides integrating MEMS mirrors to miniaturize confocal laser-scanning microscopes, an alternative approach is to use fibre-scanning confocal endoscopes [[3],[4],[5],[6],[7]]. In these confocal imaging systems, the laser light is coupled into a single fiber used both for illumination and as the pinhole for confocal detection. To achieve fiber tip scanning, the most common approaches are coupling the fiber to piezo-electric tube (PZT tube), or attaching the fiber to a tuning fork [[32],[33],[34],[35],[36]]. In the first case, the fiber is passed through a tubular piezo-electric actuator and is fixed with adhesive as shown in figure 2.4 a). In general, two pairs of electrodes on the outer surface of the PZT tube are used for actuation. By applying amplitude-modulated sine and cosine waveforms with frequencies close to the mechanical resonant frequency of the fiber cantilever, a spiral scanning pattern is achieved. Though robust and allowing an convenient position calibration, one drawback is the relatively small fiber tip deflection. With a PZT

2 Background: Fluorescence Confocal Laser-Scanning Microscopy

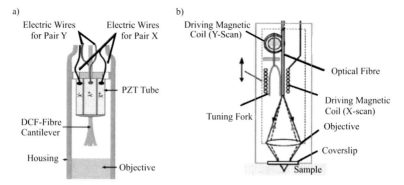

Fig. 2.4: Fiber scanning techniques based on a) PZT scanning per [37] and b) tuning fork scanning per [38]

tube length of 8.2 mm, a maximum circular field of view with a diameter of 220 µm is achieved [34]. As a result, the application of PZT fiber scanning is a good solution for endoscopic applications where a small field of view is sufficient. For larger fields of view, other techniques are necessary. The second, well-known approach of fiber tip scanning allowing a larger field of view is based on a miniaturized tuning fork, where the distal end of the fiber is attached to one arm [[35],[36]]. Using a magnetic coil driver, a 700 Hz resonant oscillation is induced in the tuning fork to provide a fast x-axis scan, while a second coil forces the entire tuning fork to pivot in the y-axis at a slower scanning rate of 1 - 2 Hz [35]. The reported maximal field of view so far is limited to 475 µm × 475 µm, for a lateral resolution of 0.7 µm [4]. In contrast to the PZT and MEMS technologies, this technique has already been implemented as a commercially available product called Optiscan Five and is sold by OptiScan Pty Ltd (Victoria, Australia). In summary, commercially available confocal laser-scanning microscopes so far are mainly based on galvanometric scan mirrors. High and stable image acquisition rates are achieved with these mirrors which even surpass those achievable with 2D MEMS mirrors and fiber scanning devices. However, the size of the galvanometric mirrors is large compared to those alternative scanning techniques. As a result, these techniques are more appropriate for applications where the microscope is required to be small and compact. For dermatology applications, either a fiber scanning approach or integrating a MEMS mirror are the obvious choices. The first endoscopes based on tuning-fork scanning recently entered the market. However, the field of view for fiber scanning devices is thus far restricted to 475 µm × 475 µm. Here MEMS mirrors show a clear advantage by offering scan angles up to ±15° in each direction. The correspondingly larger field of view achieved is especially important for dermatologic applications.

2.2 Promising F-LSM Applications in Dermatology

In dermatologic diagnostics, the usage of fluorescence confocal laser-scanning microscopy (F-LSM) as a technique for imaging and analysis of subsurface details in human skin was first presented by Swindle et al. in 2002 [39]. In the following years, Lademann et al. (University Medical School in Berlin) in various studies showed F-LSM to be a promising imaging technique in dermatologic diagnostics [[40],[41]]. Among other applications, the monitoring of epidermal wound healing, the evaluation of inflammatory and neoplastic skin disorders, and the investigation of potential effects of topically applied drugs are prominent examples [[42],[43]]. Something they all have in common is that the diagnosis is performed in vivo and that all information about the different epidermal layers gathered with the F-LSM is available within seconds. Neither painful biopsies nor long waiting periods are necessary.

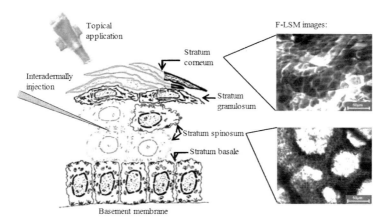

Fig. 2.5: The combination of intradermally and topically applied fluorescein (left) allows F-LSM imaging of different skin layers ([44]). Depicted on the right are images of the stratum corneum and the stratum spinosum taken by Lademann et. al. [42], Scale bars: 50 µm

2 Background: Fluorescence Confocal Laser-Scanning Microscopy

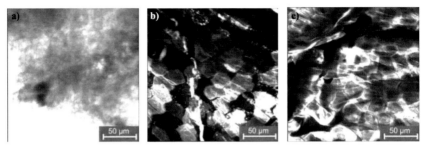

Fig. 2.6: F-LSM images of the typical epidermal wound healing process taken by Lademann et al. [42](a) Phase 1, immediately after the removal of the epidermis, wound exudation appears as a bright homogeneous cellular area due to accumulating fluorescein. (b)Phase 2, first clusters of corneocytes are found (c) Phase 3, several layers of polygonal corneocytes can be found throughout the entire wound site.

The only preparation needed for F-LSM measurements is the initial application of a fluorescent contrast agent to label the skin structure, as depicted in figure 2.5. For subsurface investigations, intra-epidermal injection is necessary, as the outermost layer of the skin acts as a diffusion barrier [45]. Since 3-4 million people in Germany suffer from a rising trend of chronic wounds, evaluation and monitoring of epidermal wound healing is a matter of particular interest [46]. In addition, it has thus far not been possible to quantitatively evaluate this process with established diagnostic procedures. So far, no quantitative evaluation has been performed in clinical practice. The gold standard for wound healing monitoring is to measure the epidermal barrier function via the trans-epidermal water loss (TEWL) [47]. Here, the amount of water evaporated through the skin is measured with a highly sensitive electrode system. Depending on the degree and area of tissue damage, the TEWL varies from high values reflecting an impairment of the water barrier in cases of severe damage, to lower values for healed wounds and thus an intact dermal barrier function [[48],[49]]. However, a disadvantage is that the TEWL data are easily influenced by external factors such as temperature, humidity, and patient perspiration, and by topically applied substances, all of which result in misleading TEWL values [50]. A quantitative comparison of different treatment procedures is therefore difficult to achieve [[40],[51]]. In contrast, F-LSM imaging can subdivide the wound healing process into three defined healing phases. Phase one is characterized by inflammation and wound exudation; in phase two, the formation of the first layer of corneocytes is visible; the third phase describes the conclusion of tissue repair, where the skin has re-established its barrier function [42]. In figure 2.6 F-LSM images showing the three phases. In contrast to TEWL measurements, F-LSM-based imaging of re-epithilialization is affected neither by treatments such as application of

2.2 Promising F-LSM Applications in Dermatology

wound ointment, nor by other factors such as perspiration or wound exudation. Thus, F-LSM not only offers an objective, quantitative, and non-invasive means of monitoring and evaluating the healing process in epidermal wounds , but also allows comparison of different treatment procedures. The following target specifications for the first F-LSM system layout are derived from this application in dermatologic diagnostics. For this first

Tab. 2.1: Target Specifications

Parameter	Symbol	Value	Unit
$\lambda_{\text{Illumination}}$	$\lambda_{\text{Illum.}}$	488	nm
$\lambda_{\text{Fluorescence}}$	λ_{f}	519	nm
Lateral Resolution	$\Delta x, \Delta y$	2	µm
Axial Resolution	Δz	4	µm
Field of View		480×480	µm
Size		Easily Portable	

design, the size of the field of view is chosen to be 480 µm × 480 µm in order to provide functional verification and to illustrate that the MEMS-based approach can offer fields of view larger than offered by the currently available endoscopes (475 µm × 475 µm), as discussed in section 2.1. The lateral and axial target resolution are chosen to be the same as the resolution offered by the endoscope (Stratum, Optiscan Ltd., Melbourne, Australia) used by Lademann et. al. in his studies [52]. The laser wavelength is selected in accordance with the fluorescent dye fluorescein used in the studies mentioned above. Further information on available dyes for dermatologic applications is given in the following section and their impact on the microscope layout is discussed.

2.3 Impact of Fluorescent Dyes on Microscopy

Fluorescence is a form of luminescence. It is a process whereby a certain molecule is excited by the absorption of a photon and returns back to the ground state by emitting a photon. This process is illustrated in figure 2.7. A molecule with singlet ground state is excited by a photon of energy $E = h\nu_A$ which is supplied by an external light source

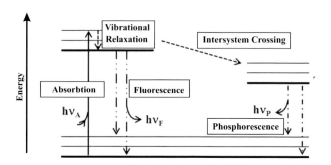

Fig. 2.7: One form of a Jablonski diagram showing the processes involved in the decay of a molecule with a ground state and excited state

such as a laser. As a result, the molecule's electrons change from their ground state to one of many vibrational levels in an excited electronic singlet state. This excited state of the molecule exists for a finite time (10^{-10} s $-$ 10^{-8} s) during which internal conversions and vibrational relaxations take place [53]. Once the molecule arrives at the lowest vibrational level of an excited singlet state, a relaxation back to the ground state can occur. This process is called fluorescence and results in the emission of a photon with energy $E = h\nu_f$ [53]. Due to energy losses during internal conversions and vibrational relaxations mentioned above, the photon emitted possesses a lower energy than the one absorbed. The change in photon energy produces a shift of the fluorescence spectrum to a longer wavelength relative to the absorption spectrum and is referred to as the Stokes Shift. Besides fluorescence, other relaxation processes can occur such as phosphorescence, for example. In contrast to fluorescence, phosphorescence occurs with a time delay exceeding 10^{-8} s [53]. The transition probabilities are much lower for these processes than for fluorescence emission and consequently do not affect the fluorescence measurement itself.

2.3 Impact of Fluorescent Dyes on Microscopy

Tab. 2.2: Fluorophores for human use in medicine (sources: [55],[56],[57],[58],[59])

Fluorophore	Abs. [nm]	Em. [nm]	Phototoxic	Selected Applications
Protoporphyrin (PpIX)	407	635	yes	Photodynamic Therapy
5-Aminolevulinic acid (5-ALA)	407	635	yes	Photodynamic Therapy
Indocyanine green (ICG)	800	830	no	Angiography, Liver function test
Sodium fluorescein	490	519	no	Fluorescence angiography
5-Aminofluorescein human serum albumin (AFL-HSA)	497	523	no	Fluorescence-guided surgery of malignant brain tumors

2.3.1 Fluorescence Contrast Agents in Medicine

Despite the fact that a huge range of fluorophores are used in biological research, only a few are approved for use in humans. In Germany, fluorophores must be approved by the Federal Institute for Drugs and Medical Devices (BfArM) in accordance with the German Medicines Act (Arzneimittelgesetz). The BfArM is an independent higher federal authority within the portfolio of the Federal Ministry of Health [54]. Its mission is to prevent health risks to the population arising from medicinal products and medical devices. In the course of the licensing procedure, the BfArM reviews the proof of efficacy, safety, and adequate pharmaceutical quality of the finished medicinal product. This process necessitates several clinical trials with the fluorophore. Due to these high standards and the time-consuming approval procedure, there are only a few fluorophores approved so far for human use. A list of these fluorophores and their current field of application is given in Tab. 2.2. Obviously, only the two fluorophores indocyanine green (ICG) and sodium fluorescein can be considered for dermatological applications, since AFL-HSF is still in phase I/II of the clinical trials and the remaining two are photosensitizers used in photodynamic therapy [[60],[55],[58],[59]]. The goal of this therapy is to destroy tissue locally by exposing the photosensitizers to light of a certain wavelength. This leads to the production of singlet oxygen, which is very reactive and destroys the surrounding tissue [[59],[61]]. Thus, the latter two fluorophores are beyond the scope of this application. The remaining two fluorophores, indocyanine green and sodium fluorescein, are both well-suited, but fluorescein has been selected in this thesis due to the fact that fluorescein was successfully used for a whole range of dermatologic studies of the skin in the past [[39] [40] [62] [63]]. As shown in figure 2.8, the maximum absorption

2 Background: Fluorescence Confocal Laser-Scanning Microscopy

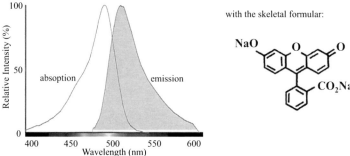

Fig. 2.8: Absorption and emission spectra from fluorescein [64] and its corresponding chemical structure (right)[65]

and emission wavelengths are 490 nm and 519 nm, respectively. It is therefore advantageous to choose a laser wavelength close to 490 nm, since the ability of the fluorophore to absorb light in this region is maximal and thus it is most effectively excited.

2.3.2 Influence of Fluorescein on the Laser Illumination

Besides the maximum absorption and emission wavelengths, further additional characteristics of the fluorophore are important for quantitative assessment of the required laser power and detector parameters. With information about the maximum possible fluorescence photon emission rate F_{max}, the absorption cross-section σ that conveys the probability of an incoming photon being absorbed, and the probability QY that an excited fluorophore will emit a fluorescence photon, it is possible to estimate the required incoming photon flux I_{sat} and corresponding laser power P_{Laser} before saturation takes place [26]:

$$F_{\text{max}} = I_{\text{sat}} \cdot QY \cdot \sigma \tag{2.1}$$

$$P_{\text{Laser}} = I_{\text{sat}} \cdot E_{\text{Photon}} \cdot A \tag{2.2}$$

Here E_{Photon} is the energy of the emitted photon and A is the focal spot area. In this case, the fluorophore is excited so rapidly that no molecules are left in the ground state and the emission is limited by the excited-state lifetime. Hence, the emission is said to be saturated and a further increase of the laser power will not lead to any higher output of fluorescence photons.

This maximum number of photons a single fluorophore is able to emit per second is restricted by its excited-state lifetime, referred to as the fluorescence lifetime τ [66]. The

2.3 Impact of Fluorescent Dyes on Microscopy

duration of this lifetime is the average time a fluorophore spends in the excited state and depends on the fluorophore and ambient conditions. Fluorescein itself exhibits a moderate lifetime of 4.1 ns that leads to a maximum emission rate F_{max} of [66]:

$$F_{max} = 1/\tau = 2.44 \cdot 10^8 \text{ Photons}/\text{s}. \tag{2.3}$$

The probability that an excited fluorophore decays by emitting a fluorescence photon and not by any other decay process, such as phosphorescence, is called the fluorescence quantum yield QY. This fluorescence QY depends on the fluorophore, with values ranging from zero for non-fluorescent compounds to one for a 100% efficiency [66]. For fluorescein, the QY is a comparatively high 0.93 [66]. The probability of a photon being absorbed by the fluorophore is described by the molecular cross section σ of the molecule. Directly related to it is the decadic molar extinction coefficient ϵ of the fluorophore. With the knowledge of the fluorophore concentration c in moles/liter and the molecular cross section σ, the extinction coefficient is derived through the two Lambert Beer law representations:

$$T = 10^{-\epsilon c l} = e^{-\sigma n l} \tag{2.4}$$

The Lambert Beer law predicts a logarithmic dependence between the transmission T of light through a substance with the cross section of light absorption by a single molecule, the number density of particles n and the optical path length l of the light. Instead of the cross section, it is generally the extinction coefficient ϵ of the specific fluorophore which is given to specify its absorption behavior. For fluorescein, the extinction coefficient at 488 nm is $80,000$ M^{-1}cm^{-1} at a pH of 7 [66], which leads to an optical cross-section σ per molecule of:

$$\sigma = \frac{\epsilon \cdot \ln(10) \cdot c}{n} \tag{2.5}$$

$$= \frac{10^3 \cdot \ln(10)}{N_A} \cdot \epsilon \tag{2.6}$$

$$= 3.06 \cdot 10^{-16} \text{ cm}^2 \tag{2.7}$$

where n is the number density of the fluorescein molecules given by

$$n = N_A \cdot c \tag{2.8}$$

with N_A being Avogadro's Number. For comparison, the average diameter of a fluorescein molecule is 6.9 nm, which implies an absorption rate of around 10% for all incoming photons with a wavelength of 488 nm. By setting these values for ϵ, QY, and F_{max} into

2 Background: Fluorescence Confocal Laser-Scanning Microscopy

Tab. 2.3: Characteristic Properties of Fluorophores (sources: [66],[67],[26])

Property	Definition	Fluorescein
excitation spectrum	Spectrum where the number of emitted fluorescence photons is plotted versus the excitation wavelength	
Stokes Shift	Shift in the wavelength between the absorption and emission spectra caused by energy losses	
Extinction coefficient (ϵ)	Molar absorptivity of the analyte at the absorption maximum wavelength	$80,000\ \text{M}^{-1}\text{cm}^{-1}$
Quantum Yield (QY)	Ratio of the number of emitted fluorescence photons to the number of absorbed photons	0.93
Fluorescence Lifetime (τ)	Average time the fluorophore spend in the excited state	4.1 ns
Photobleaching	Irreversible photochemical reaction that modifies the fluorophore in such a way that it no longer fluoresces	

relation to each other, it is then possible to estimate the excitation photon flux I needed before reaching saturation and to estimate the corresponding laser power:

$$I = \frac{F}{\sigma \cdot QY} = 8.57 \cdot 10^{23}\ \text{cm}^{-2} \cdot \text{s}^{-1}. \tag{2.9}$$

In combination with an assumed resolution in lateral direction of $\omega_0 = 1\ \mu\text{m}$, which refers to the $1/e^2$ intensity beam radius in the focal area, the necessary laser power is then given by:

$$P_{\text{Laser}} = I_{\text{max}} \cdot E_{\text{Photon}} \cdot \pi \cdot \omega_0^2 = 2.7\ \text{mW} \tag{2.10}$$

A higher laser power in the focal spot area does not lead to an increase of the number of emitted photons; however, taking photon losses in the illumination path into account, it should be chosen higher. Not withstanding, high laser powers will accelerate the destruction of excited fluorophores through photobleaching. An overview of these

2.3 Impact of Fluorescent Dyes on Microscopy

different characteristic properties of fluorophores and corresponding values for fluorescein are given in table 2.3. The number of photons available for detection and thus the necessary detector specifications depends in addition on the fluorophore concentration in the sample and the optics utilized. These and related parameters will be discussed in more detail in the following chapter.

To summarize, the choice of a suitable fluorophore, the knowledge of its main characteristics as well as the fluorescence processes in general is important to avoid unexpected consequences. This is especially true for applications in the field of medical diagnostics where only a few fluorophores are available and thus even fewer lasers are appropriate. For dermatologic diagnostics, fluorescein is a well-known and commonly utilized fluorophore. The final choice of fluorophore not only imposes certain requirements on the laser specifications, it also affects the overall optics significantly. The degree of correction for aberration of both the illumination and emission wavelengths of the fluorophore determine to a great extent overall resolution, contrast and brightness of the microscope.

2 Background: Fluorescence Confocal Laser-Scanning Microscopy

2.4 Introduction to the Wavefront Aberration Theory

As mentioned above, optical aberration determine to a great extent overall resolution, contrast and brightness of the microscope. It is thus important to know their effect on the image quality as well as the relation between them and the figures-of-merit provided in the design program for the interpretation of the simulation results. Aberrations are deviations of the ideal, diffraction-limited performance of an optical system. These deviations from ideal image formation are the result of defects, misalignments of components, and aberrations inherent in the optical system design. Aberrations of construction include incorrect shapes, thickness, and improper mounting and spacing of the different lens elements, for example. These are taken into account during the design process by tolerance analysis, which will be discussed in the next chapter. In the present chapter, the different aberrations inherent in the optical system, their effect on the image quality, and criteria to evaluate them will be introduced. In paraxial systems, none of these aberrations are present due to the fact that rays with small angles of incidence are assumed. In this case, the sine function in Snell's Law is approximated by $\sin\theta = \theta$, the first order term in the Taylor series expansion:

$$n_1 \sin\theta_1 = n_2 \sin\theta_2 \quad (2.11)$$

$$\sin\theta = \theta - \frac{\theta^3}{3!} + \frac{\theta^5}{5!} - \frac{\theta^7}{7!} + ... \quad (2.12)$$

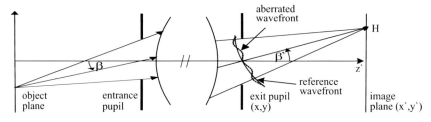

Fig. 2.9: Schematic of an optical imaging system with an exaggerated illustration of an aberrated wavefront for an off-axis point object

Here θ_1 is the angle of incidence and θ_2 the angle of refraction at the interface between two homogeneous isotropic media of different indices of refraction n_1, n_2. However, this paraxial approximation becomes unsatisfactory for larger angles of θ. Larger angles occur in particular for high numerical apertures and large field angles, as present in confocal scanning systems. As a result, the wavefront at the exit pupil of real optical systems is no longer perfectly spherical, but is instead aberrated. This deviation of the aberrated wavefront from the reference spherical wave wavefront emerging from the exit

2.4 Introduction to the Wavefront Aberration Theory

pupil of the system is described by the wave aberration function $W(\rho.\theta)$ [[68], [69]]. In general, this aberration function is expanded as a power series, where the exit pupil coordinates or the aperture coordinates x, y are expressed in polar coordinates[[68], [69]]:

$$W(H,\rho,\theta) = a_1\rho^2 + a_2H\rho\cos\theta + b_1\rho^4 + b_2H\rho3\cos\theta + ... \quad (2.13)$$
$$... + b_3H^2\rho^2\cos\theta^2 + b_4H^2\rho^2 + b_5H^3\rho\cos\theta + O(6) \quad (2.14)$$

with

$$x = \rho\cos\theta, y = \rho\sin\theta \quad (2.15)$$

Here, ρ is the normalized pupil height, θ is the azimuth angle, and H the normalized image height or field coordinate. The first two terms in eq. 5.7 with the coefficients a_1 and a_2 correspond to defocus and tilt. They originate from longitudinal and transverse shift of the center of the reference sphere, respectively. The next five terms are all fourth-power and account for the five monochromatic aberrations of axially symmetrical optical systems. The wave aberrations corresponding to the coefficients b_1 to b_5 are called spherical aberration, coma, astigmatism, field curvature, and distortion [68]. They are commonly referred to as the Seidel, or third-order aberrations, whereas "third-order" refers to the power sum when expressed as transverse ray aberrations [69]. These formed the basis of Seidel's investigations in 1856 and are the partial derivatives of the wave aberration with respect to the pupil coordinate. In the following the different wave aberrations are shortly introduced. These aberrations are discussed in optical literature e.g. by H. Gross et al. [69] and V. N. Mahajan [68].

Spherical aberration $\Delta W = b_1\rho^4$
Spherical aberration is independent of the field coordinate. Light rays from outer regions of the lens are brought to focus closer to the vertex than those rays passing through the system close to the optical axis. Spherical aberration exhibits a fourth-order dependence on the aperture coordinate and varies with the lens shape, lens orientation, and the refractive index of the material. Spherical aberration is an issue especially in systems where a large aperture is combined with a lens of small focal length. In general, a positive, converging lens or surface will lead to under-corrected spherical aberration; a negative lens or divergent surface will do the reverse.

Coma $\Delta W = b_2H\rho3\cos\theta$
Coma can be defined as the variation of the magnification with the aperture and appears as a comet-like flare in the image plane. Rays from different zones of the aperture

2 Background: Fluorescence Confocal Laser-Scanning Microscopy

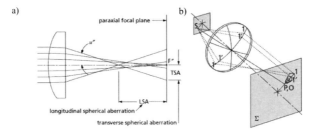

Fig. 2.10: Illustration of a) spherical aberration [70] and b) coma [70]

stop cross the image plane at different heights from the optical axis. It shows a cubic dependence on the aperture coordinate ρ. On axis ($H = 0$), no coma is present.

Astigmatism $\Delta W = b_3 H^2 \rho^2 \cos \theta^2$
Astigmatism occurs when rays in the tangential and sagittal planes focus at different distances from the lens. In the presence of astigmatism, a spot diagram shows the image of a point as an ellipse in either one of the focal planes. The best compromise position is located between both foci, where the image blur is a circle. The amount of astigmatism depends quadratically on the field coordinate H. Due to this dependence, it even occurs for small beam diameters or apertures.

Field curvature $\Delta W = b_4 H^2 \rho^2$
Field curvature is a field-dependent longitudinal focal shift. Positive lens elements usually have inward curving fields. In the absence of astigmatism, this curved field is called a Petzval surface. Its radius is directly correlated to the curvature and the refractive index of a lens in the optical system. In a system consisting of several lenses, the Petzval curvature is the sum of all contributions from the different lenses.

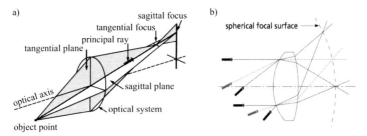

Fig. 2.11: Illustration of a) astigmatism and b) field curvature [70]

2.4 Introduction to the Wavefront Aberration Theory

Distortion $\Delta W = b_5 H^3 \rho \cos\theta$
Distortion is a field dependent variation of the magnification that leads to a distorted image. If the off-axis magnification is larger than the on-axis magnification, the resulting distortion is called pin cushion distortion (see figure 2.12). In contrast to all other primary aberrations, each point is sharply focused in the image plane. Distortion present in images can be corrected by post-processing.

Chromatic aberration
In addition to the above monochromatic aberrations, wavelength-dependent aberration can also occur. These chromatic aberrations are caused by dispersion of light by the lens material, due to the variation of its refractive index n with the wavelength of light. The resulting variation of the system's focal length for different wavelengths is either called axial chromatic aberration or transversal chromatic aberration, depending on the direction in which it occurs. A measurement of the material's dispersion is given by the Abbe number or V-number (V). This number is defined as [69]:

$$V = \frac{n_d - 1}{n_F - n_C} \tag{2.16}$$

where n_d, n_F, n_C are the refractive indices of the material at the wavelengths of the Fraunhofer d-, F- and C- spectral lines (587.6 nm, 486.1 nm and 656.3 nm, respectively). A correction of chromatic aberrations can be achieved by choosing lens materials with appropriate dispersion properties and by replacing singlets with lens combinations. One example of the latter is an achromatic doublet, where in general a low-dispersion ($V > 55$) positive crown-glass lens is combined with a high-dispersion ($V < 55$) negative flint-glass lens [69]. Here, crown and flint relate to the Abbe number V of the material greater than 55 or less, respectively.

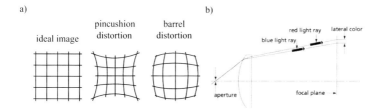

Fig. 2.12: Illustration of a) different kind of image distortions [70] and b) lateral chromatic aberration [70]

2.5 Important Figures of Merit for Image Quality Evaluations

In a real optical system, a combination of the aberrations described above is present and negatively influence the image quality. To evaluate the optical performance of the system, different figures-of-merit are used to describe the image quality. Useful means of characterization in the image plane are, among others, spot diagrams, changes in size and shape of the point spread function (PSF), and the Strehl ratio. In the exit pupil, the optical path difference (OPD) of the aberrated wavefront from the reference sphere is a common measure of the aberrations present in the system. The OPD is described by the wave aberration function in units of wavelength and is generally plotted versus position in the exit pupil for different image heights H. Directly correlated to the OPD are evaluation criteria such as the root-mean-square wavefront error W_{rms} and the peak-to-valley (P-V) value. The P-V value simply states the maximum departure of the actual wavefront from the ideal wavefront in both positive and negative directions. Due to the fact that this value does not provide any information about the area over which the departure occurs, it is only suitable for wavefront errors with small changes in slope. For larger changes in slope, the RMS wave-front error is better suited because it provides an average level of the error across the pupil and is defined as [69]:

$$W_{rms} = \sqrt{\langle W^2 \rangle - \langle W \rangle^2} \qquad (2.17)$$

Another approach for characterizing the effects of aberrations and diffraction is to evaluate the change in size and shape of an image formed by the optical system. For the image of a point source, the intensity distribution in the image plane is given by the point spread function PSF, which is defined as the quare of the absolute value of the Fourier transform of the pupil function $P(x,y)$ in the exit pupil [71]:

$$PSF(x', y') = |FT[P(x,y)]|^2 \qquad (2.18)$$

with P(x,y) in the exit pupil defined by

$$P(x,y) = p(x,y) \cdot e^{-i\frac{2\pi}{\lambda}W(x,y)} \qquad (2.19)$$

Here, p(x,y) represents the shape, size and transmission of the exit pupil, while W(x,y) is the wave aberration function in units of wavelength. Directly correlated to the PSF is the Strehl ratio, which is defined as the ratio of the peak intensity of the PSF in the presence of aberrations to the peak intensity of an aberration-free optical system [69]:

$$\text{Strehl Ratio} = \frac{\text{PSF}^{\text{real}}(0,0)}{\text{PSF}^{\text{ideal}}(0,0)} \qquad (2.20)$$

2.5 Important Figures of Merit for Image Quality Evaluations

The Strehl ratio as a description of system quality only makes sense in systems with low aberrations and Strehl numbers above 0.6, since all the information about the three-dimensional PSF profile is reduced to only one number [69]. Strehl ratios > 0.8 are commonly referred to as diffraction-limited system performance based on the Rayleigh criterion and correspond closely to a P-V value of $\lambda/4$ [69]. For systems with aberrations much larger than the diffraction limit, the spot diagram or geometrical PSF are more meaningful. In these diagrams, all ray intersections relative to the paraxial image point are plotted for rays from a single point source traced though the entrance pupil. For a homogeneous ray distribution in the entrance pupil, the density of spots in the image plane corresponds to the energy density. This is the reason why the spot diagram is also called the geometrical PSF. However, it is a purely geometrical estimate of the image blur produced by the system aberrations and diffraction is not taken into account.

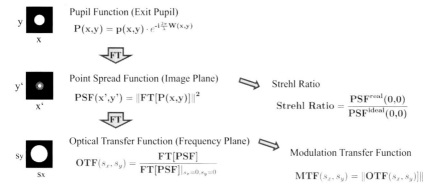

Fig. 2.13: Relation between the Pupil function, PSF, Strehl ratio, MTF and OTF of an imaging system

It is thus not a correct figure-of-merit for diffraction-limited systems. Nevertheless, for systems where the OPD is larger than several wavelength, these diagrams allow types of aberrations present to be identified by their characteristic appearance in the image plane, such as coma for example. To estimate the field curvature, chromatic, and astigmatic effects quantitatively, design programs frequently offer plots of different field curves vs. field height. Other frequently used criteria for system performance are the Optical Transfer Function OTF and the Modulation Transfer Function MTF [69]. MTF is the modulus of the OTF and measures how well a lens preserves contrast from the subject to the image. The relations between the PSF, Strehl Ratio, MTF and OTF are depicted in figure 2.13. These figures-of-merit are provided in the design program and are used to evaluate an optical system during the design process and to locate appropriate components. Furthermore it is possible to define appropriate merit functions during the

25

2 Background: Fluorescence Confocal Laser-Scanning Microscopy

design process. By defining such a merit function, the evaluation of the design can be automated and defined variables optimized by the design program itself to improve the overall imaging performance.

3 MEMS Mirror Based F-LSM Design

In this chapter the optical system design of an initial large field of view F-LSM set-up based on a MEMS mirror will be presented to give functional verification of the MEMS-based approach. The first part starts with general considerations in the optical design process, where implications and requirements resulting from the integration of the MEMS mirror on the remaining optical components are discussed and correlated figures of merit for the microscope performance are outlined. In the following an in detail discussion of the simulation results of optical system design followed by a tolerance analysis will be given. The impact of mounting and fabrication tolerances on the optical performance is discussed. The effect of dynamic deformations of the scanning mirror on the optical performance are evaluated. Finally the occurrence of distortions when a laser beam is projected by a dual-axis MEMS mirror onto a image plane image is discussed and correction possibilities introduced.

3.1 MEMS mirror based Implications on the Microscope Specifications

The layout of the here presented microscope is comparable to a traditional confocal fluorescence laser scanning microscope as introduced in the last chapter, yet it features a single dual-axis MEMS mirror as a core component. A particular property of this gimbal mount mirror developed at the Fraunhofer IPMS is, that both axis are scanned simultaneously, which is referred to as cardanic scanning. Hereby no additional relay optics are necessary to image one mirror onto the other as it is commonly done when galvanometric mirrors are applied. The employed dual-axis MEMS mirror in this microscope features a large mirror size D_{mirror} of 2 mm diameter while maintaining a fast scanning speed at resonant frequencies of 199 Hz and 1.337 kHz at the slow and fast axis, respectively. The chosen mirror for the F-LSM system deflects the laser beam with a maximum mirror deflection angle of ϕ of $\pm 5°$ in both directions and a Lissajous pattern is scanned. Due do the favorable properties of this mirror as there are its vibration insensitive layout with large scan angles combined with fast scanning frequencies it is possible to achieve a robust F-LSM layout with a large field of view. However, the integration of this dual-axis MEMS mirror considerably affects the overall system layout. The accompanied challenges and implications on the remaining optical elements included in this design will be

3 MEMS Mirror Based F-LSM Design

outlined in the following and improtant figures of merit for the microscope performance, as there is the lateral and axial resolution as well as the number of fluorescence photons reaching the detector will be discussed and set into relation with the MEMS mirror and the chosen components. Taking the target application in dermatology into account the remaining elements are chosen to be mainly off-the-shelf components to demonstrate a cost effective layout.

3.1.1 Implications on the Illumination Optics

For illumination a 488 nm fiber coupled laser with a collimator at the distal end of the fiber is used. This collimated laser light is directed onto the MEMS mirror by a dichroic mirror which used for the separation of the illumination and the fluorescence light in the detection path as depicted in figure 3.1. This dichroic mirror preferentially reflects the laser wavelength and transmits the fluorescence light. To ensure Gaussian illumination at the sample the beam diameter of the laser has to be chosen with respect to the limiting system aperture. This is due to the fact, that the determining factor for the appearance of changes in the Gaussian laser beam profile is the ratio of the aperture-to-beam-radius $T = r_{aperture}/r_{beam}$ [72]. The more the Gaussian laser beam is truncated the greater is the energy spread and loss in the focal area [72]. Changes in the characteristics even occur for weak diffraction in the far field for an aperture-to-beam radius in the range of [73]:

$$1.6 > r_{aperture}/r_{beam}. \qquad (3.1)$$

Here r_{beam} is given by the $1/e^2$ intensity beam radius r_0 in the entrance aperture and the aperture size itself is defined by the mirror. In these cases the beam still looks like a Gaussian beam in the far field with side lobes appearing [73]. In this system the entrance aperture coincides with the mirror plane and r_{beam} is given by the $1/e^2$ intensity beam radius r_0. The aperture itself is an ellipsoide due enclosed angle of the incoming light beam with the surface normal of the mirror in y direction. With a minimal enclosed angle of $\phi = 30°$ due to geometrical constraints this leads to an effective mirror diameter and thus decreased aperture diameter in this direction to:

$$D_{eff} = \cos(30°) \cdot D_{mirror} = 1.73 \text{ mm} \qquad (3.2)$$

Based on the effective aperture diameter and to ensure Gaussian illumination while minimizing the following relay and magnification optics as will be discussed later on, the laser beam diameter has been chosen to be 1.1 mm which is equivalent to an aperture-to-beam-radius of:

3.1 MEMS mirror based Implications on the Microscope Specifications

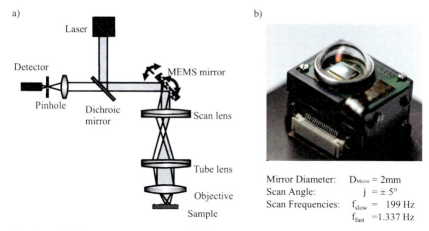

Fig. 3.1: a) Schematic optical train of the fluorescence confocal laser scanning microscope based on a dual-axis MEMS mirror, b) photograph of the dual-axis MEMS mirror module with integrated electronic interface (Fraunhofer IPMS)

$$T = \frac{r_\text{eff}}{r_0} = 1.6, \qquad (3.3)$$

in x-direction and T=0.55 in y-direction. To calculate the spot diameter in presence of truncation an adjusted aperture shape factor k_{FWHM} is required [72]. The corresponding k_{FWHM} function which permits the calculation for the above stated beam truncation ratio is given by [74]:

$$k_{FWHM,x} = 1.036 - \frac{0.058}{(\frac{1}{T})} + \frac{0.156}{T^2} = 1.33 \qquad (3.4)$$

$$k_{FWHM,y} = 1.036 - \frac{0.058}{(\frac{1}{T})} + \frac{0.156}{T^2} = 1.45 \qquad (3.5)$$

With this k-factor the smallest detectable scan angles in both scan directions x and y are given by:

$$\Delta\phi_x = k_{1/e^2} \cdot \lambda \cdot \frac{1}{D_\text{eff}} \qquad (3.6)$$

$$\Delta\phi_y = k_{1/e^2} \cdot \lambda \cdot \frac{1}{D_\text{mirror}} \qquad (3.7)$$

3 MEMS Mirror Based F-LSM Design

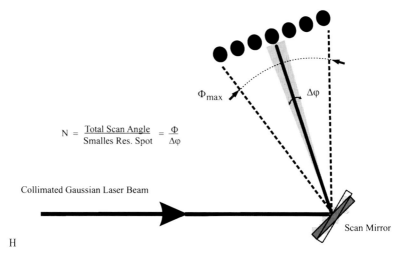

$$N = \frac{\text{Total Scan Angle}}{\text{Smalles Res. Spot}} = \frac{\Phi}{\Delta\varphi}$$

H

Fig. 3.2: Total number of resolvable spots of a scanner system. With a laser beam reflected by a mirror with diameter D a full angular scan Φ is generated. The total number of resolvable spots N is determined by the diffracted angle $\Delta\varphi$ emanating from the aperture D in combination of the full angular scan angle. $\Delta\varphi$ can be expressed as $\sin(\Delta\varphi) \approx \Delta\varphi = k \cdot \frac{\lambda}{D}$ for $\Delta\varphi$ being small

Here the difference in the x and y direction originates from the angle enclosed by the incoming light beam with the surface normal of the mirror in x direction. This leads to an effective mirror diameter or aperture diameter D_{eff} in x direction. In combination with the total optical scan angle Θ of the MEMS mirror total number N_Θ of resolvable spots in both directions is determined:

$$N_{\Theta_x} = \frac{\Theta \cdot D_{\text{eff}}}{k_{FWHM} \cdot \lambda} = 931 \tag{3.8}$$

$$N_{\Theta_y} = \frac{\Theta \cdot D_{\text{mirror}}}{k_{FWHM} \cdot \lambda} = 857 \tag{3.9}$$

This is proportional to the so called ΘD-product [[75],[76]] and inversely proportional to the light wavelength $\lambda_{\text{em}} = 520$ nm and the aperture shape factor k. Even though the scan angle Θ may be increased by the following optics the number of resolvable spots keeps unchanged. Thus, this number is solely determined by the size and the scan angle of the MEMS mirror in combination with the laser beam diameter.

To increase this number of resolvable spots it is in consequence necessary to either increase the size of the mirror plate and the beam diameter accordingly or to use larger

3.1 MEMS mirror based Implications on the Microscope Specifications

scan angles. This relation is referred as resolution invariant I_N and is expressed by :

$$I_N = \Theta D = \Theta' D' \tag{3.10}$$

in analogy to the Lagrange invariant, which states that the product of the field, refractive index and aperture for rays propagating through an optical system are constant [75]. In the equation above the factor $\frac{1}{k \cdot \lambda}$ is omitted since it is assumed to be the same for ΘD and $\Theta' D'$. Based on these results, necessary specification of the imaging optics will be discussed in the following. Here, important parameter are once again the scan angle and the incoming laser beam diameter which finally lead to the maximum achievable field of view for a defined optical resolution.

3.1.2 Requirements on the Imaging Optics and the Z-Shifter

The main objective of the imaging optics downstream of the scan mirror is to transform the angular movement of the laser beam leaving the MEMS mirror to a lateral movement of the focal spot in the specimen plane, and to ensure a Numerical Aperture (NA) appropriate for optical sectioning and fluorescence measurements. Considering the fact that most microscope objective lenses are corrected telecentrically, pupil matching is accomplished by placing the mirror such that its pivot point is at the center of a conjugate plane to the objective entrance pupil. To image the mirror onto this entrance pupil, which coincides with the entrance aperture in the back focal plane of the objective, intermediate relay optics are necessary. Keplerian telescope optics are used consisting of a scan and tube lens spaced the sum of their focal lengths apart. This arrangement is double telecentric, since both entrance and exit pupil lie at infinity. In addition to imaging the mirror in the entrance pupil of the objective, it also expands the laser beam (see figure 3.3). The resulting angular movement of the beam filling the entrance pupil of the objective is transformed into a lateral movement of the focus in the specimen plane. An adequate objective NA for the present application is 0.4. This NA not only determines the resolution but also defines the fraction of the fluorescence light which is collected. In consequence, the objective, which is used to focus the beam onto the sample, needs to exhibit at least an NA of 0.4. Further important requirements in the field of biological applications involving 3D measurement are reasonable working distance, chromatic correction, and high transmission. Because of the difference in excitation and emission wavelength of the fluorescence specimen, the color correction is especially important. Considering these requirements, the decision fell on a Nikon CFI Plan Apochromat 20×0.75 with a focal length $f = 10$ mm and a working distance of 1 mm. With the present NA of 0.4, which is smaller than the achievable NA of 0.75 of

3 MEMS Mirror Based F-LSM Design

the objective, the resulting spot radius ω_0 and Rayleigh range z_r are

$$\omega_{0x} = 0.5 \cdot k_{1/e^2} \cdot \lambda_{\text{ex}} \cdot f/\# = 0.43 \, \mu\text{m} \tag{3.11}$$

$$z_{rx} = \frac{\pi}{\lambda_{\text{ex}}} \cdot \omega_0^2 = 1.2 \, \mu\text{m} \tag{3.12}$$

$$\omega_{0y} = 0.5 \cdot k_{1/e^2} \cdot \lambda_{\text{ex}} \cdot f/\# = 0.47 \, \mu\text{m} \tag{3.13}$$

$$z_{ry} = \frac{\pi}{\lambda_{\text{ex}}} \cdot \omega_0^2 = 1.4 \, \mu\text{m} \tag{3.14}$$

in this case

$$k_{1/e^2} = 1.6449 + \frac{0.6460}{(\frac{1}{T} - 0.2816)^{1.821}} - \frac{0.5320}{(\frac{1}{T} - 0.2816)^{1.891}} \tag{3.15}$$

is applied due to the fact that ω_0 in general refers to the $1/e^2$ laser beam diameter [70]. To achieve the above calculated Rayleigh range, the laser beam entering the objective needs to have a diameter of

$$D = 2 \cdot NA \cdot f = 7 \, \text{mm} \tag{3.16}$$

which results effectively in an under-filling of the objective pupil and impedes further beam clipping. As a consequence, the intermediate optics have to provide a beam-expansion ratio of at least $M = 6.4$. That way, the illumination beam expands from a diameter of 1.1 mm at the mirror site to 7 mm in the entrance pupil of the objective. To realize this magnification of the beam diameter, the intermediate optics are arranged like those of a Keplerian telescope, forming an image of the mirror in the entrance pupil of the objective. A particular property of this telecentric arrangement is that any chief ray of the scanned beam lies parallel to the optical axis in image space (see figure 3.3). In addition, lateral movement of the beam in the objective's entrance pupil is avoided, which would otherwise result in beam clipping. Based on this telescope principle, the magnification M results from the ratio of the focal lengths of the scan lens $f_{\text{scan}} = 20$ mm located close to the scanning mirror and of the tube lens $f_{\text{tube}} = 140$ mm next to the objective. Note that these are common focal lengths of lenses available in catalogues, keeping the cost low. In addition, the relatively short focal length of the scan lens is advantageous, since it not only provides enough space for mounting, but also minimizes the optical path throughout the intermediate optics. The resulting magnification is:

$$M = \frac{f_{\text{tube}}}{f_{\text{scan}}} = 7 \tag{3.17}$$

3.1 MEMS mirror based Implications on the Microscope Specifications

Directly correlated to this magnification of the beam diameter is the reduction of the scan angle in the entrance pupil of the objective by the same factor:

$$\frac{D}{D_{\text{laser}}} = \frac{\Theta}{\Theta_{\text{obj}}} = M \qquad (3.18)$$

This relation is depicted in figure 3.3 and refers to the Lagrange invariant, which was introduced in the context of the ΘD product. The ratio of entrance and exit beam diameters and the ratio of entrance and exit scan angles always follow this inverse relationship, as can easily be proven. With this relationship and the specifications of the objective, the resulting field of view has a diameter of:

$$d_{\text{im}} = 2 \cdot f \cdot \tan 2 \cdot \phi / M = 500 \text{ µm} \qquad (3.19)$$

where ϕ denotes the mechanical mirror deflection angle of 5° and the optical resolution is given by the Housten criterium to [77]:

$$\Delta x = 0.5 \cdot k_{FWHM,x} \cdot \lambda_{\text{ex}} \cdot f/\# = 0.27 \text{ µm} \qquad (3.20)$$
$$\Delta y = 0.5 \cdot k_{FWHM,y} \cdot \lambda_{\text{ex}} \cdot f/\# = 0.29 \text{ µm} \qquad (3.21)$$

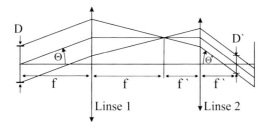

Fig. 3.3: Illustration of the resolution invariant $I_N = \Theta D = \Theta' D'$. The scanned angle $\Theta/2$ from aperture D is transfered to the output angle $\Theta'/2$ at output aperture D' so that $\Theta'/\Theta = D/D'$ (from [75])

For the acquisition of images at different defined depths inside the fluorescent specimen, additional axial movement of the focal plane in the specimen has to be possible. There are basically two ways to achieve this movement while keeping the object stationary: a) an axial movement of the objective lens, or b) a movement of the focus by the intermediate optics [24]. The latter can be ruled out, since changing the focus position by moving the intermediate optics would require involve distances that were far too large. This is due to the fact that the axial magnification M_z is the square of the lateral

33

3 MEMS Mirror Based F-LSM Design

magnification M_{lat}:

$$M_z = M_{lat}^2 \tag{3.22}$$

Thus, in our case with the focal length of the objective $f = 10$ mm and $f_{tube} = 140$ mm the lateral magnification is $M_{lat} = 14$. In consequence a focus shift of 10 µm in the object plane would lead to a shift of

$$\Delta z_{\text{int}} = \Delta z_{\text{Obj}} \cdot M_{lat}^2 = 1.96 \text{ mm} \tag{3.23}$$

in the intermediate image plane. Hence, the method of choice is to use a piezoelectric z-shifter to move the objective. This arrangement ensures excellent, backlash-free operation over a range of 100 µm with position repeatability in the z-direction of ±10 nm.

3.1.3 Fluorescence Detection Path

The confocal character of the F-LSM is due to the special layout of the detection path, whereby a pinhole prevents out-of-focus light from reaching the detector. For detection, a fraction of the emitted fluorescence light is collected by the objective, descanned along the original illumination path, and finally separated from the excitation light by a dichroic mirror. The identical optical path for scanning and descanning is possible due to the fact that the angular velocity of the scanning mirrors together with the short optical paths is very small relative to the speed of light, with the result that the fluorescent light passing the objective follows the identical path as the original excitation beam. Subsequent to the dichroic mirror, additional filters are used for further elimination of the excitation light before the fluorescence light is focused onto the pinhole. By placing this pinhole in a conjugate focal plane to the specimen plane, lateral and axial discrimination is achieved due to the fact that out-of-focus light in the axial direction will be focused either in front or behind the pinhole and out-of-focus light in the focal plane will miss the pinhole in the lateral directions. Besides limiting the detection to fluorescence emission occurring within the focal volume, the additional pinhole optimizes the image contrast by suppressing stray light. The schematic principle of such a confocal set-up is depicted in figure 3.4. The size of the pinhole is usually set to be equal to a diameter of $0.7 - 1.5$ of the Airy disk diameter (1 Airy unit) [78]. The Airy disk is defined as the size of the central region of the diffraction pattern, measured from the disk center to the center of the trough of the first minimum. According to the Rayleigh Criterion, the minimal resolvable distance of two diffraction patterns is given by the radius of the Airy disk. In consequence, $0.7 - 1.5$ Airy units correspond to a diffraction-limited or nearly diffraction-limited spot on the plane of the pinhole [78]. Further reduction of the pinhole size does not considerably improve the optical sectioning capability [79]. With a given magnification M of the relay system and a lens of $f_{pin} = 40$ mm focal length for

3.1 MEMS mirror based Implications on the Microscope Specifications

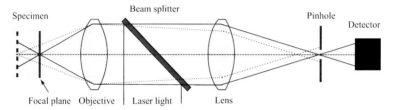

Fig. 3.4: Confocal detection principle from [67]

focusing the fluorescence light onto the pinhole, the NA at the pinhole is given by

$$NA_{\text{pin}} = \frac{NA_{\text{Obj}} \cdot f_{\text{Obj}}}{M \cdot f_{\text{pin}}} = 0.014 \qquad (3.24)$$

and yields an Airy Unit (AU) of

$$D_{\text{Airy}} = 1.22 \cdot \frac{\lambda_{\text{em}}}{NA_{\text{pin}}} \qquad (3.25)$$

for maximum fluorescence emission spectra around 520 nm. Here, the NA_{Obj} is only 0.4 and not the achievable NA of 0.75 of the objective. This is due to the fact that the MEMS mirror acts as the limiting aperture. Thus, choosing a pinhole of 50 μm diameter relates to a diameter of 1.04 AU and a throughput of around 80% of the light in the central maximum [80]. The total number of photons reaching the detector is an important selection criterion in choosing the most appropriate one. To estimate this, two aspects have to be considered: a) the number of fluorescence photons available for detection, given by the number of emitted photons in the focal volume, and b) the number of remaining photons finally reaching the detector.

Estimation of Emitted Fluorescence Photons in the Focal Volume

The number of fluorescence photons available for detection depends on four factors: a) the applied laser power, b) the focal volume, c) the dye concentration and d) the pixel illumination time. For a sinusoidal motion of the resonant MEMS mirror, the pixel illumination time at the center of the scan is given by:

$$t = \frac{1}{\pi \cdot f_{\text{fast}} \cdot 1024} = 2.33 \cdot 10^{-7} \text{ s} \qquad (3.26)$$

with f_{fast} denoting the fast axis scanning frequency of 1337 Hz. Assuming a laser power greater than 2.7 mW for saturated fluorescence emission as introduced in chapter 2, each

3 MEMS Mirror Based F-LSM Design

fluorescence molecule can emit

$$N_{\text{Photons/Molecule}} = 1/\tau \cdot t = 56 \tag{3.27}$$

Photons. Here, $\tau = 4.1$ ns denotes the fluorescence lifetime of fluorescein. With the given 1/e spot radius in lateral and axial direction of 0.43 µm and 1.2 µm respectively, the focal volume is thus given by:

$$V = \frac{4}{3} \cdot \pi \cdot r_{\text{lat}}^2 \cdot r_{\text{axial}} \tag{3.28}$$

$$V = \frac{4}{3} \cdot \pi \cdot 0.43 \text{ µm}^2 \cdot 1.2 \text{ µm} \tag{3.29}$$

$$V = 9.4 \cdot 10^{-16} \text{ l} \tag{3.30}$$

Considering a moderate dye concentration of $c_{\text{dye}} = 10$ µMol/l, the number of dye molecules within the focal volume is estimated to be:

$$N_{\text{dye}} = V \cdot c_{\text{dye}} \cdot N_A = 5.680 \tag{3.31}$$

where N_A denotes the Avogadro constant. This leads to a minimum number of emitted photons per pixel of:

$$N_{\text{Photon}} = N_{\text{dye}} \cdot F \cdot t_{\text{scan}} = 3.23 \cdot 10^5 \tag{3.32}$$

Despite this relatively high number, only a fraction of the emitted fluorescence photons reach the detection path, since emission occurs isotropically. With an NA of 0.4, corresponding to an apex angle $\alpha = 2 \cdot sin^{-1} NA = 47.16°$, only photons emitted within the solid angle of

$$\Omega = 2 \cdot \pi \cdot (1 - \cos\frac{\alpha}{2}) = 0.52 \tag{3.33}$$

can enter the objective and hence the detection path. These are solely

$$\frac{\Omega}{\Omega_{total}} = \frac{1 - cos\frac{\alpha}{2}}{2 \cdot \pi} = 4\% \tag{3.34}$$

of all emitted fluorescence photons in the focal volume, which is equivalent to around 12,900 photons per pixel available for detection.

In the detection path itself, further photon losses occur which reduce the number of photons finally reaching the detector. These additional photon losses originate from scattering and absorption effects of the pinhole and preceding optics, filters, and mirrors. These photon losses occur on all diffractive and refractive surfaces. The assumed intensity losses at the different elements in the detection path are summarized in ta-

3.1 MEMS mirror based Implications on the Microscope Specifications

Tab. 3.1: Intensity losses at the different elements in the detection path

Element	Loss (Percent)	Remainder (Percent)
Sample emission		100
Collection by effective NA of 0.4	96	4
Relay optic	10	3.6
MEMS mirror	15	3.1
Dichroic mirror	5	2.9
Filter	20	2.3
Pinhole	20	1.9
Percentage of emitted photons reaching the detector		1.9

ble 3.1. In total, only 1.9% of the emitted photons are available for detection, which corresponds to around 6,000 photons from the originally 323,000 emitted fluorescence photons per pixel. Thus, these low light levels in the range of 5 nW arriving at the detector necessitate selection of a highly sensitive detector that also has high bandwidth > 5 MHz due to the short pixel dwell time of 2.33 µs and has a high level of sensitivity to a continuous flux of varying light intensity at a wavelengths of 520 nm. To evaluate and compare the suitability of available detectors, several performance parameters need to be considered, such as the responsivity S, the photon detection efficiency PDE, the signal-to-noise ratio S/N, and the noise equivalent power NEP . Additional criteria are the detectors' bandwidth and their integrated preamplifiers that convert the signal current into a signal voltage. The final voltage/power response is given by the responsivity S multiplied by the preamplifier gain $G_{\text{preamplifier}}$:

$$\text{voltage/power}_{\text{response}} = S \cdot G_{\text{preamplifier}} \quad (3.35)$$

where the responsivity S determines the available output signal of a detector for a given optical input signal:

$$S = PDE \cdot G_{\text{internal}} \frac{e}{h\nu} \quad (3.36)$$

Here, G_{internal} is the internal gain of the photodetector, $e = 1.602 \cdot 10^{-19}$ C is the electron charge, $h\nu$ the photon energy, and PDE the photo detection efficiency in %, which defines the probability of detecting an incident photon at the detector. Taking these considerations into account, a detector from SensL having an active area of 1 mm in diameter and a bandwidth of > 20 MHz has been chosen. This detector has a responsivity on the order of 50 $^{kA}/w$ and a preamplifier gain of 2,000 V/A, leading to a voltage/power response of around 0.1 $^{V}/_{nW}$. This allows a fluorescence input signal of around 1 nW to be converted to a signal voltage of 100 mV. In addition, a C-mount adapter is available that facilitates attachment of the detector on the tube system, which is very advantageous for pinhole detector coupling while avoiding the influence of stray light.

3.2 Intermediate Relay Optics - Performance Simulations with Zemax

To develop a portable and cost effective F-LSM, constraints such as minimal space requirement and the utilization of off-the-shelf components were applied during the design process. In order to achieve an optimal arrangement, the imaging optics shown in figure 3.5 was simulated by means of commercially available optical design software using figures-of-merits discussed in chapter 2.3 for evaluation. The rays in the simulation were traced forward from infinity at the mirror plane through the intermediate optics to the exit pupil of the objective using different field angles and wavelengths.

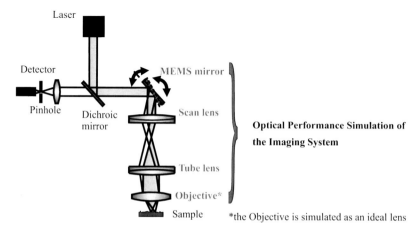

Fig. 3.5: Schematic of the optical train of the F-LSM based on a dual-axis MEMS mirror where the imaging optic simulated in the design process are highlighted

Off-the-shelf lenses with focal lengths of 20 mm and 140 mm were chosen for the scan and tube lens, respectively. The primary selection criteria for these lenses were a small focal length of the scan lens to ensure minimal space requirements, and good chromatic correction to account for the wavelength difference of 30 nm between the fluorescence excitation and emission light. Chromatic correction is crucial due to the fact that both the excitation and emission beams pass through the relay lens. As a result, both lenses are achromats, with a Hasting triplet for the scan lens and a doublet for the tube lens. An additional advantage of using achromats to avoid dispersion effects, is that they are known to have less spherical aberration. With a relatively short focal length of 20 mm, the scan lens not only provides enough space for mounting but also minimizes the optical path throughout the intermediate optics. An ideal lens with 10 mm focal length was

3 MEMS Mirror Based F-LSM Design

assumed for the objective in the design, since microscope objectives are usually very well-corrected lenses. The simulation was performed for different field angles at two wavelengths of 488 nm and 520 nm, corresponding to the laser and fluorescence emission wavelengths, respectively. The optical design layout and the corresponding lens data are depicted in figure 3.6, table 3.2 and table 3.3.

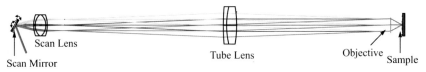

Fig. 3.6: Optical layout with rays traced for three different deflections of the dual-axis MEMS mirror. The objective lens is modeled as an ideal lens

Tab. 3.2: F-LSM System Specifications

Parameter	Symbol	Value	Unit
$\lambda_{\text{Illumination}}$	$\lambda_{\text{Illum.}}$	488	nm
$\lambda_{\text{Fluorescence}}$	λ_f	519	nm
Laser Beam Diameter	D_{laser}	1.1	mm
Lateral Resolution	$\Delta x, \Delta y$	0.8	μm
Field of View, $\varnothing$		500	μm

Tab. 3.3: Components

Mirror	Scan Lens	Tube Lens	Objective
2D MEMS Mirror	Hasting Triplet	Achromat	CFI Plan Apo 20x
(Fraunhofer IPMS)	(Edmund Optics)	(Qioptiq)	(Nikon)
$D_{\text{eff}} = 1.73$ mm	$f_{\text{scan}} = 20$ mm	$f_{\text{tube}} = 140$ mm	$f = 10$ mm
$\Theta = \pm 5°$	$d_{\text{scan}} = 12$ mm	$d_{\text{tube}} = 22.4$ mm	NA=0.75

The Strehl ratio was used as a figure-of-merit for the final optimization and analysis stages. According to the Rayleigh criterion, diffraction-limited performance is achieved for a Strehl number larger than 0.8, which corresponds to a Peak-to-Valley value of the wave aberration smaller than $\lambda/4$ [69]. As shown in figure 3.7 b), nearly diffraction-limited performance is present over the full field. In addition, the corresponding PSF in figure 3.7 a) shows that the spot size is around 1.2 μm for the maximum angle of 10°. This demonstrates the capability of the system to image cell nuclei with a diameter of 6 μm. This level of optical performance makes the set-up an adequate tool for cell

3.2 Intermediate Relay Optics - Performance Simulations with Zemax

imaging in biology. The Optical Path Difference is plotted in figure 3.8 for different field angles to display the kind of aberrations present in the system more completely. Coma is obviously the dominant aberration in this geometry at larger field angles. The main reason is the remote stop position given by the arrangement of the MEMS mirror in front of the scan lens. This eliminates the symmetry of the optical system about the principal ray, which would help to reduce coma, distortion, and lateral chromatic aberration Because the Hasting Triplet chosen is meant for applications as magnifying lens, it is designed to reduce pincushion distortion, lateral chromatic aberration, and spherical aberration. Unfortunately, simultaneous correction of coma and spherical aberration is not possible in this case when a reasonably flat field is required. Thus, we have to tolerate a higher amount of coma due to the correction of spherical aberration, even though correction in our case wouldn't have been as necessary. The reason is the relatively small beam diameter in comparison to the focal length of the scan lens, which already tends to hold spherical aberration to a reasonable value.

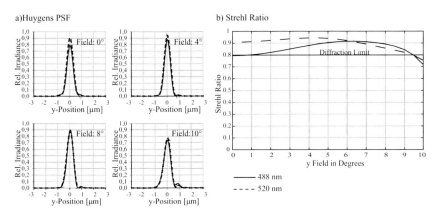

Fig. 3.7: Predicted performance of the optical design in figure 3.6 a) Huygens PSF for four different field angles b) Strehl Ratio as a function of the field angle

In addition, the system possess some moderate field curvature of 1.8 μm as illustrated in figure 3.8 b), which does not affect the image quality considerably due to the effect of the pinhole. The pinhole rejects out-of-focus light and automatically images the surface having the best focus. In the case of biological samples, their preparation already influences the position and elongation to a much greater extent than the field curvature. Hence aberrations such as defocus and field curvature are nearly eliminated. In total, aberrations restrict the maximum scanning angle to ±10°. The maximum spot size for

41

3 MEMS Mirror Based F-LSM Design

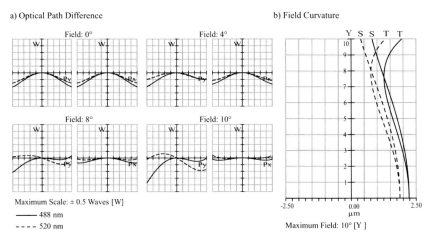

Fig. 3.8: a) Wave aberration at the exit pupil of the objective for four different field angles b) Field curvature in tangential and sagittal directions

a field angle of 10° is around 1.2 µm in diameter. The effective optical angle, which corresponds to the illumination of the corner of the field of view, given by $\sqrt{2} \cdot 10°$. The design is therefore expected to meet the target specifications and to provide adequate performance over the whole circular field-of-view of 0.5 mm in diameter for the intended application of cell imaging. The final specifications of the system are listed in table 3.4.

Tab. 3.4: Optical System Specifications

Type	Radius (mm)	Thickness (mm)	Glass	Semi-Diameter
Distance to Mirror		16.35		0.60
	14.95	1.10	N-F2	6.30
Hasting Triplet	7.35	7.98	N-BK7	6.30
(Edmund Optics)	-7.35	1.10	N-F2	6.30
	-14.95			6.30
Air		154.70		
	61.75	6.00	N-BK7	12.70
Achromat (Qioptiq)	-87.22	2.00	N-SF5	12.70
	-177.83			12.70
Air		126.58		
Objective (f=10mm)		0.00		7.50
Image Distance		10.00		7.50

3.3 Analysis of Assembly Tolerances

Before starting with the assembly it is important to prove how the optical system reacts to tolerances. The knowledge about whether a system is sensitive to changes in the parameters such as glass or air thickness, decenter, tilt etc. decides on the feasibility to built up a system with a performance close to the predicted one by the simulations. Here a correct estimation of the variations which come along with the values defining the optical system is crucial to achieve results which are neither to optimistic nor to pessimistic. The general tolerances applied for the microscope imaging optics are presented in 3.5. They can be divided into manufacturing tolerances given by the lens tolerances and mounting tolerances referred as positioning tolerances. For the refractive index and the Abbe number of the lenses the tolerances are given by the standard delivery quality for fine annealed glass from Schott AG. This refers to a tolerance of ± 0.0005 and $\pm 0.5\%$ for the refractive index and the abbe number respectively. For manufacturing tolerances such as surface tilt and decentration fairly standard tolerances where assumed with $0.025°$ and $33\mu m$ respectively and for the lens thickness and the radius variation information where obtained from data sheets. Hereby the radius tolerance was not given directly but instead the maximum change in the focal length. With this data for the focal length tolerance of 1% for the hasting triplet and 0.5% for the achromat the corresponding radius tolerance ist possible to estimate. With the comparatively low lens thickness of the Hastings triplet and the achromat lens, it is possible to use the thin lens equations for this estimation. The relation between the focal length f and the radius r of a thin double convex lens is given by:

$$f = 0.5 \frac{n_{glass}}{n_{glass} - n_{air}} \cdot r \qquad (3.37)$$

With the propagation of uncertainty this leads to following change in the focal length:

$$\Delta f = \left| \frac{\delta f}{\delta r} \Delta r \right| + \left| \frac{\delta f}{\delta n_{glass}} \Delta n_{glass} \right| \qquad (3.38)$$

$$= 0.5 \cdot \left| \frac{n_{air}}{n_{glass} - n_{air}} \Delta r \right| + 0.5 \cdot \left| \frac{-1}{(n_{glass} - 1)^2} \cdot r \cdot \Delta n_{glass} \right| \qquad (3.39)$$

By assuming $n_{air} \approx 1$, $\Delta n_{glass} \approx 0.0005$ and a refractive index $n_{glass} \approx 1.5$ the second term is negligible. In consequence the equation can be rewritten and leads to a nearly linear relation between the change of the radius and the focal lengths:

$$\Delta f \approx \frac{\delta f}{\delta r} \Delta r \qquad (3.40)$$

3 MEMS Mirror Based F-LSM Design

with

$$\Delta f \approx 0.5 \frac{n_{air}}{n_{glass} - n_{air}} \cdot r \cdot P\% \qquad (3.41)$$

$$\Delta f = f \cdot P\% \qquad (3.42)$$

Here P denotes the tolerance of the radius as percentage. Thus a focal length tolerance of 1% for the hasting triplet and 0.5% for the achromat refer to approximately 1% and 0.5% tolerance on the lens radius, respectively. Final simulations with the optical simulation program showed good agreement of these estimated results with the simulated results. Changing the radii of curvature of all lens surfaces from the Hastings triplet by -1% and then by $+1\%$ of the nominal 20.03mm focal lengths produced 19.85 mm and 20.2 mm respectively. This corresponds to a change of the focal lengths of around $\pm 1\%$ (with -0.9% and $+0.9\%$, respectively.

The mounting tolerances such as decentration and tilt and axial positioning accuracy, which are listed in table 3.5 as positioning tolerance - thickness, have been estimated in correspondence with the choosen mount. Here it was taken into account that decentrations of the scan lens does affect the overall system performance to much higher extend than the tube lens due to its shorter focal lengths. This leads to higher variations of the magnification with the aperture so that decentration will result in on-axial coma. Considering this relation an x-y translation stage was assumed for the scan lens mounting whereas no additional centering option was assumed for the tube lens. The corresponding tolerances for decentration from the global optical axis are listed in table 3.5.

Due to the fact that collimation is fairly easy to test in the final set-up the distance between the scan and tube lens was set as variable during the tolerancing process to account for these adjustments during the assembly process. Besides this compensator a so called automatic compensators is present which adjusts without direct input from the user. In this design this is the image distance since confocal microscopes automatically image the slice of best focus in a 3D object even though it can mean it is a curved one. For final evaluation of the system, tolerancing was performed with a Monte Carlo statistical analyses, where a defined number of random system arrangements was created. By this method all the parameters defining the optical system are randomly varied within the set tolerance ranges according to the assumption that the parameters follow the same normal distribution within the defined range. Hereby a statistical study is carried out, where all tolerances are simulated simultaneously.

3.3 Analysis of Assembly Tolerances

Tab. 3.5: Tolerance Assumptions for the Monte Carlo Analysis

Tolerances on	Value (±)	Unit
Radius of Curvature		
Scan Lens	1	%
Scan Lens	0.5	%
Lens Element Thickness		
Scan Lens	30	µm
Tube Lens	50	µm
Element Distance		
Eye Relief	200	µm
Tube Lens - Objective	500	µm
Element Decenter		
Scan Lens	50	µm
Tube Lens	200	µm
Element Tilt		
Scan Lens Element 1,2	0.05	°
Tube Lens	0.1	°
Surface Decenter		
All Lens Surfaces	30	µm
Surface Tilt		
All Lens Surfaces	0.033	°
Index		
All Glasses	0.0005	
Abbe Number		
All Glasses	0.8	%
Distance Compensator 1	Scan Lens - Tube Lens	
Distance Compensator 2	Image Distance	

In the next step the resulting systems are optimized by using the prior defined compensators as variables for optimization according the tolerance criteria. Here, the RMS wavefront error was chosen as tolerance criteria. Whether a near diffraction limited system performance over the whole field of view is realistic to achieve was arbitrarily defined at the 80% Monte Carlo level as Strehl number higher >0.7. During this analysis 50 systems with random combinations for the given set of manufacturing and mounting tolerances are tested.

3 MEMS Mirror Based F-LSM Design

In summary the tolerance analysis revealed in more than 80% the systems are close to a diffraction limited performance for all scan angles as depicted in figure 3.9. By the sensitivity tolerance analysis provided by the optical design program the worst offenders are listed, refering to the most sensitive parameters. Here, the distance between the mirror and the scan lens is the most sensitive parameter followed by manufacturing tolerances of the scan lens itself, such as surface decentration and radius tolerances. In consequence care has to be taken in the adjustment of the scan lens.

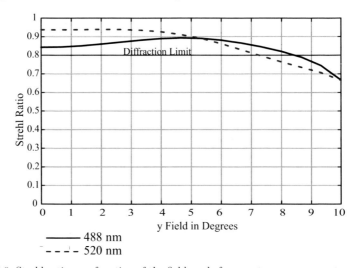

Fig. 3.9: Strehl ratio as a function of the field angle for a system arrangement generated by the Monte Carlo analysis with an optical performance worse than 80% of all generated system arrangements

3.4 Dynamic Deformations of the MEMS Mirror

Besides the above stated tolerances all surface imperfections of folding mirrors and dynamic deformations of the scanning mirror influence the imaging performance negatively. To keep these additional performance reductions low both folding mirrors where chosen to have surface flatnesses of at least $\lambda/10$. In contrast to them, the surface flatness of the dual-axis MEMS mirror is not critical, since it is in the range of a few nm in correspondence to the original SOI substrate.

However, high acceleration forces during the scanning process lead to dynamic deformations, which are in consequence the highest at the turning points. These deformations δ_{mirror} originating from inertial moments are of particular interest since they scale with the mirror size D, thickness t, scan angle Θ and scanning frequency f_{scan} as follows [81]:

$$\delta_{\text{mirror}} \propto \frac{D_{\text{mirror}}^5 \cdot f_{\text{scan}}^2 \cdot \Theta_{\text{scan}}}{t} \tag{3.43}$$

In the present case of a comparatively large mirror size of 2 mm, a scan frequency of $f_{fast} = 1337$ Hz and a large maximum scan angle of around 7.5° it was therefore important to calculate the maximum plate deformation to estimate the effect on the optical system performance. To calculate the deformed shape of the mirror for an angle of 7.5° at the fast axis, Finite Element Analysis (FEA) was used. These simulations where performed by a colleague from Fraunhofer IPMS [82].

Mirror deformation for a scan angle of 7.5° with corresponding Zernike polynomial coefficients

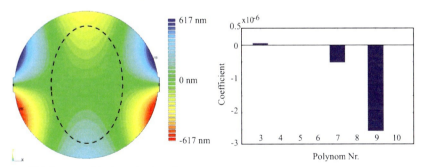

----- 1/e² diameter of the incoming laser beam
RMS = 20.3 nm, PV = 123.5 nm

Fig. 3.10: Dynamic MEMS mirror deformation with the expression of the deformation by the first ten Zernike polynomial coefficients for a scan angle of 7.5°(FEA simulation performed by a colleague from Fraunhofer IPMS [82])

The result of this calculation is shown in figure 3.10 with the corresponding description of the mirror surface deformation by the first ten Zernike polynomial coefficients. Zernike polynomials present an alternative description of the classical wave-front aberrations inherent in an optical system and are often used for data fitting and the expression of deformed optical surfaces such as mirrors. In the case here, the notation defined in R. Noll, "Zernike polynomials and atmospheric turbulence"[83] was used. In this notation the Zernike polynomials are orthonormal, with the first terms being related to the traditional Seidel aberrations [[80], [83]]. The calculated dynamic MEMS mirror deformation for a deflection angle of 7.5 ° depicted in figure 3.10 shows a maximum P-V height difference of the mirror surface of 123.47 nm and a RMS value of 20.3 nm. Such a height difference in the center of the mirror would be highly unfavorable, however these large deformations only occur at the edges. Within the region of $1/e^2$ laser beam radius the maximum deformations are significantly lower with a P-V value of around 36 nm. This correlates to a deformation of less than $\lambda/10$ in the blue and consequently does not affect the image quality considerably. Dynamic deformations, originating from the slow frame axis with $f_{slow} = 199\,\text{Hz}$ are more than a hundred times lower and thus negligible due to the fact that these deformations scale quadratically with the scan frequency. In summary, dynamic deformations in the central region of the mirror surface are even for large scan angles low in comparison to inherent aberrations in the optical layout. For comparison, inherent optical aberrations lead to P-V values of around 100 nm for scan angles of 5°, as shown in the previous section. Finally, it should be noted that it is important to take the different scanning frequencies for both mirror axis and the dependence of the dynamic mirror deformations on them into account to chose an appropriate mirror to incoming laser beam orientation. For applications where even lower deformations are necessary, this is possible to achieve by an improved spring design, where for example additional springs are placed around the mirror plate [84].

3.5 Image Distortion - Causes and Correction Methods

For laser projection systems where a laser beam is projected by a dual-axis MEMS mirror onto a image plane image distortion occurs. The origin of this beam path distortion is mainly the scanning mechanism itself and not by remaining lens aberrations. These distortions have been discussed and mathematically described for a few special mirror-plane arrangements in previous works by Yajun Li and Scholles et al. [85],[86],[87]]. In this thesis, a mathematical model based on the vector analysis approach from Scholles et al. will be introduced first to describe the microscope scanning system for the case when the center of the reflection of the incident laser beam onto the mirror coincides with the center of mirror rotation. Based on these results, an algorithm is derived which allows for distortion correction of the images in post-processing. The algorithm is then demonstrated with some microscope images. In the following this model is discussed for more general cases where the image plane is tilted and the center of reflection does not coincide with the center of mirror rotation.

Figure 3.11 shows the arrangement of the laser scanning system on which the following descriptions are based. The resulting geometrical distortion of a projected image on the screen is illustrated.

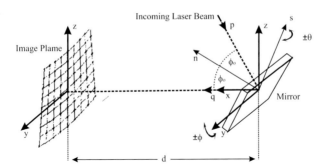

Fig. 3.11: Optical schematic of the dual-axis MEMS mirror scanning system

Main parameters affecting the degree of pincushion distortion are the following:

- angle of incidence ϕ_0 of the laser beam onto the mirror surface
- scan angles ϕ and θ of the mirror
- distance d along the x-axis between rotation mid-point of the mirror and the image plane
- tilt of the image plane

3 MEMS Mirror Based F-LSM Design

- distance between the point where the incoming laser beam hits the mirror and the center of rotation from the mirror

For this mathematical model, the influence of the last parameter, which accounts for decentering, is neglected. It is assumed in this model that the incident laser beam hits the mirror at its centre point of rotation. The basis chosen for the coordinate system is depicted in figure 3.11, with its origin in the center of the laser beam reflection at the mirror. The x-axis is chosen such that the laser beam reflected by the mirror in its rest position will coincide with the x-axis. The direction of the incident laser beam is described by vector $\vec{p}$ and the direction of the laser beam reflected by the mirror surface towards the projection screen by vector $\vec{q}$. By taking into consideration that $\vec{q_0}$ coincides with the x-axis and that the incoming laser beam denoted by vector $\vec{p}$ encloses the angle $\phi_0 = 30\,°$ with the normal vector $\vec{n}$ of the mirror as introduced in chapter 3.1, the vector $\vec{n_0}$ normal to the surface of the mirror in its rest position is in consequence located in the xz-plane enclosing the angle $\phi_0 = 30\,°$ with the x-axis and the incident laser beam $\vec{p}$. $\vec{n_0}$ and $\vec{p}$ are given by:

$$\vec{n_0} = \begin{pmatrix} n_1 \\ n_2 \\ n_3 \end{pmatrix} = \begin{pmatrix} \frac{1}{2}\sqrt{3} \\ 0 \\ \frac{1}{2} \end{pmatrix} \tag{3.44}$$

$$\vec{p} = \begin{pmatrix} p_1 \\ p_2 \\ p_3 \end{pmatrix} = \begin{pmatrix} \cos 2\phi_0 \\ 0 \\ +sin 2\phi_0 \end{pmatrix} = \begin{pmatrix} \frac{1}{2} \\ 0 \\ \frac{1}{2}\sqrt{3} \end{pmatrix} \tag{3.45}$$

When the mirror is in its rest position, the first axis of rotation of the mirror coincides with the y-axis. The second, fixed axis of rotation in the direction of $\vec{s}$, is depicted in figure 3.11. The angle between this fixed axis of rotation and the z-axis is given by $90° - \phi_0 = 60°$, leading to:

$$\vec{s} = \begin{pmatrix} -\cos 2 \cdot \phi_0 \\ 0 \\ \sin 2 \cdot \phi_0 \end{pmatrix} \tag{3.46}$$

Taking the mirror rotation about these two axis into account, the resulting vector normal to the mirror $\vec{n}$ is derived to be:

$$\vec{n} = D_s \cdot D_y \cdot \vec{n_0} \tag{3.47}$$

$$\vec{n} = \begin{pmatrix} \cos(\phi_0 + \phi) \cdot (s_1^2(1 - \cos\theta) + \cos\theta) - \sin(\phi_0 + \phi) \cdot s_1 s_3 (1 - \cos\theta) \\ \cos(\phi_0 + \phi) \cdot s_3 \sin\theta + \sin(\phi_0 + \phi) \cdot s_1 \sin\theta \\ \cos(\phi_0 + \phi) \cdot s_1 s_3 (1 - \cos\theta) - \sin(\phi_0 + \phi) \cdot (s_3^2(1 - \cos\theta) + \cos\theta) \end{pmatrix} \tag{3.48}$$

3.5 Image Distortion - Causes and Correction Methods

with the following matrices D_y and D_s used for the rotation by an angle of ϕ and θ about both axes of rotation:

$$D_y = \begin{pmatrix} cos\phi & 0 & sin\phi \\ 0 & 1 & 0 \\ -sin\phi & 0 & cos\phi \end{pmatrix} \quad (3.49)$$

$$D_s = \begin{pmatrix} s_1^2(1-\cos\theta) + \cos\theta & s_1 s_2(1-\cos\theta) - s_3 \sin\theta & s_1 s_3(1-\cos\theta) + s_2 \sin\theta \\ s_1 s_2(1-\cos\theta) + s_3 \sin\theta & s_2^2(1-\cos\theta) + \cos\theta & s_3 s_2(1-\cos\theta) - s_1 \sin\theta \\ s_3 s_1(1-\cos\theta) - s_2 \sin\theta & s_3 s_2(1-\cos\theta) + s_1 \sin\theta & s_3^2(1-\cos\theta) + \cos\theta \end{pmatrix} (3.50)$$

In the following, $\vec{n}$ is expressed as (n_1, n_2, n_3).

To find a general expression for $\vec{q}$ denoting the reflected laser beam, the law of reflection is used. Based on this law, both vectors $\vec{p}$ and $\vec{q}$ for the incident and reflected laser beams, respectively, include the same angle with respect to $\vec{n}$ normal to the mirror. As a result, the cross products between each of them and the vector normal have to be identical:

$$\vec{p} \times \vec{n} = \vec{q} \times \vec{n} \quad (3.51)$$

In addition to having the same cross products, the dot products, where one vector is projected onto the other, must also be identical.

$$\vec{p} \cdot \vec{n} = \vec{q} \cdot \vec{n} \quad (3.52)$$

By combining these two conditions, it is possible to derive the expression for $\vec{q}$:

$$\vec{q} = \begin{pmatrix} n_1^2 p_1 - n_2^2 p_1 - n_3^2 p_1 + 2n_1 n_2 p_2 + 2n_1 n_3 p_3 \\ n_1^2 p_2 - n_2^2 p_2 - n_3^2 p_2 + 2n_1 n_2 p_1 + 2n_2 n_3 p_3 \\ n_3^2 p_3 - n_1^2 p_3 - n_2^2 p_3 + 2n_2 n_3 p_2 + 2n_1 n_3 p_1 \end{pmatrix} \quad (3.53)$$

Further simplification of the equation is possible by considering the normalization of the $\vec{n}$:

$$n_1^2 + n_2^2 + n_3^2 = 1. \quad (3.54)$$

With the knowledge of $\vec{q}$, it is now possible to calculate the points of intersection from the reflected laser beam with the screen at a distance d from the mirror as a function of the different scan angles ϕ and θ. The image plane, which is parallel to the zy plane of

3 MEMS Mirror Based F-LSM Design

the coordinate system, is described in vector form by:

$$\vec{E} = \begin{pmatrix} d \\ 0 \\ 0 \end{pmatrix} + [\begin{pmatrix} 0 \\ 1 \\ 0 \end{pmatrix}, \begin{pmatrix} 0 \\ 0 \\ 1 \end{pmatrix}] \quad (3.55)$$

Figure 3.12 displays some images before and after application of post-processing corrections and in figure 3.13 the simulation result for a scan pattern of ±5° in both directions is shown before and after post-processing .

Fig. 3.12: Comparison of images before (left side) and after (right side) post-processing of images taken with the first F-LSM system

3.5 Image Distortion - Causes and Correction Methods

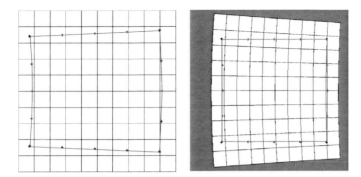

Fig. 3.13: Comparison of the simulation result for a scan pattern of ±5° in both directions before (left side) and after (right side) post-processing

These calculated points of intersection are based on the assumption that the center of the laser beam reflection and center of the mirror rotation coincide. For situation where due to a misalignment of the mirror or laser beam the laser beam position does not coincide with center of the mirror additional keystone distortion is the result.
In figure 3.14 this situation is illustrated. A misalignment of the laser beam position from the center of the mirror leads to points of intersection with the image plane which are not equidistant from one another. To correct this distortion via post-processing is possible to assume a tilted image plane along the y and z axes and then to adjust these tilt parameters until the keystone distortion in the image vanishes. In this case, only minor adjustment of the program is necessary.
To account for the different kinds of tilt, the vectors (0,1,0) and (0,0,1), which define the image plane, are first rotated α_y around the y-axis and then α_z around the z-axis using the rotation matrices D_y and D_z:

$$D_y = \begin{pmatrix} \cos(a_y) & 0 & \sin(a_y) \\ 0 & 1 & 0 \\ -\sin(a_y) & 0 & \cos(a_y) \end{pmatrix} \quad (3.56)$$

$$D_z = \begin{pmatrix} \cos(a_z) & \sin(a_z) & 0 \\ -\sin(a_z) & \cos(a_z) & 0 \\ 0 & 0 & 1 \end{pmatrix} \quad (3.57)$$

As a result, the parametric form of the plane can be written as:

$$E : x + y \cdot \underbrace{-\frac{\sin(a_z)}{\cos(a_z)}}_{b} + z \cdot \underbrace{-\frac{\sin(a_y)}{\cos(a_z) \cdot \cos(a_y)}}_{c} = d \quad (3.58)$$

3 MEMS Mirror Based F-LSM Design

The intersection points (x, y, z) of the reflected laser beam defined by $\vec{q}$ with the image plane E is then derived as:

$$x = \frac{d}{1 + b\frac{q_2}{q_1} + c\frac{q_3}{q_1}}, \qquad (3.59)$$

$$y = \frac{q_2}{q_1} \cdot x, \qquad (3.60)$$

$$z = \frac{q_3}{q_1} \cdot x \qquad (3.61)$$

In consequence misalignments of the laser beam or mirror position can be corrected by choosing the parameters accordingly.

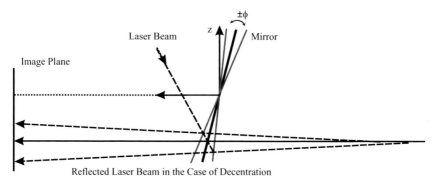

Fig. 3.14: Schematic illustration of the effect when the center of laser beam reflection does not coincide with the center of the mirror rotation. The new points of intersection are not equidistant from one another, leading to additional keystone distortion

4 F-LSM System Realization

The following section will provide functional verification of the MEMS-based laser scanning approach with a first F-LSM set-up. First, the final layout of the microscope is introduced and special issues in the assembly and adjustment process are discussed. In the second part, the optical performance will be evaluated and compared with simulation results. For the imaging performance analysis appropriate resolution targets for the axial and lateral axes are used. This evaluation is followed by wavefront measurements with a Shack-Hartmann sensor and proves the lateral resolution being limited by the calculation of the laser beam position and not the optics. Finally, images taken from several biological samples, such as images of histological probes, are presented and the specifications of the microscope system are summarized and compared with the target specifications.

4.1 F-LSM Demonstration System Layout

In figure 4.1, the principle set-up with the schematic optical train included and a photograph of the final demonstration system is presented. Here, the emphasis is on demonstration of the feasibility of achieving a large-field-of-view confocal microscope based on a MEMS mirror, while keeping the package compact and the cost low. For this reason, the microscope system is mainly composed of off-the-shelf components listed in Table 4.1 in combination with custom-designed parts and electronics developed in-house. All parts are mounted on a 220 mm × 300 mm base plate, which leads to an overall size for the F-LSM demonstrator set-up of 145 mm × 260 mm × 340 mm. This includes all components except for the fiber-coupled laser and piezo-driver for the z-shifter, which are in separate housings. To display the images taken with the microscope, an additional USB-connected laptop is used. The entire demonstration set-up with all components is depicted in figure 4.1a), while figure 4.1 b) shows a photograph of the F-LSM demonstration set-up.

4 F-LSM System Realization

a) F-LSM Demonstration System

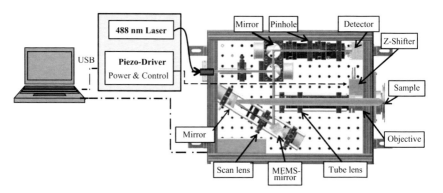

b) F-LSM Demonstration System

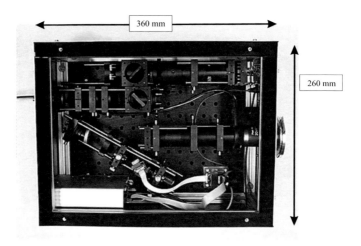

Fig. 4.1: a) Principle set-up: laptop, remotely connected power and control units and the microscope system with the schematic optical train b) Photograph of the MEMS-based F-LSM demonstration system

4.1 F-LSM Demonstration System Layout

Tab. 4.1: Major components used in the F-LSM Demonstration System Set-up

Manufacturer	Component
Coherent	488 nm, 30 mW fiber-coupled diode laser system
Schaefter+Kirchhoff	Fiber collimator with inclined coupling axis, APC connector
Optosigma	Five-axis optical mount, x-, y-, z- & tip/tilt-adjustment
Qioptiq	Microbench/Tube system components, achromat (140 mm)
Edmund Optics	Hastings triplet (20 mm)
Nikon	Objective, CFI Plan Apochromat VC 20x0.75
Physik Instrumente	Pifoc P-721 nanopositioner for objectives (100 µm of travel)
Chroma	Dichroic filter, zt 488 RDC
Semrock	Band-pass filter, HC 525/50
Sensl Technologies	Detector, SPMMini 1000 series

A fiber-coupled 488 nm laser with a collimator on the distal end is used for illumination. A collimator with an inclined coupling axis was chosen to avoid back-reflection of the laser light. To couple the 1.1 mm diameter collimated laser beam into the microscope system as described in chapter 3.1 and to ensure the correct height, position and axial orientation of the laser beam, a 5-axis mount from Optosigma was used. This is necessary due to the fact that even small tilts can lead to missing the 50 µm pinhole in the detection path that the fluorescent light is focused onto, as the pinhole can only be adjusted ±0.5 mm. For example, deviations of 0.05 ° of the incident angle from the optical axis result in missing a 50 µm pinhole in a distance of 20 cm by:

$$50 \text{ µm} - \tan(0.05°) \cdot 200 \text{ mm} = 125 \text{ µm} \quad (4.1)$$

The five-axis mount is fixed to the back side of the microscope as shown in figure 4.2 a). Including the mount inside the housing was not possible due to the fact that the optical height of this mount is 54 mm, 14 mm higher than the height of the microbench holders from Qioptiq used for all the other components. For this reason, an adapter for the five axis mount was designed and fixed on the aluminum profile of the housing to compensate for this height difference. Guide rails were used to further minimize decentration and tilt during mounting. The laser beam is then directed onto a dichroic mirror fixed on a mirror mount with axes independently adjustable via adjustment screws with 0.15 mm pitch. The laser beam is directed onto the dual-axis MEMS mirror using this dichroic mirror. To mount the dual-axis MEMS mirror an x-y mount combined with an adjustable prism holder was chosen. The mirror was fastened with an suitably designed adapter, depicted in figure 4.2 c).

4 F-LSM System Realization

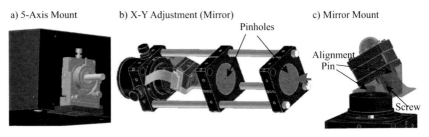

Fig. 4.2: Computer-aided design of a) the five-axis mount fastened on the back side of the microscope b) x-y adjustment set-up for the dual-axis MEMS mirror and c) the mirror mount

Alignment pins were included to ensure stable mating with tight tolerances. An initial coarse alignment was performed using two pinholes placed in front of the mirror before inserting this assembly into the microscope system. If the mirror is placed in the center of the optical axis, the only thing that can be seen through both pinholes is the mirror plate. To adjust the subsequent optics, the laser was placed in the position where the objective would be placed. The tube lens, the scanning lens, and the additional folding mirror were adjusted using pinholes and shear plates. As discussed in chapter 3, the scan lens is mounted in a x-y adjustment unit, whereas no additional x-y adjustment of the tube lens is integrated. It is important to note here that the scan lens was positioned axially by focusing the laser light onto the mirror before the tube lens was inserted. A shear plate was used for collimation testing. Such a plate acts as a shearing interferometer, where interference is observed and used to test the collimation from laser beams. Laser beams typically have a coherence length much longer than the thickness of the shear plate, so the basic condition for interference is fulfilled. To align the different mirrors, the set-up was extended using rods, a mount and pinholes. In addition to the dichroic mirror for separating the laser excitation light from the fluorescence, two additional band-pass filters rejecting 99% of the laser light were used in the detection path to ensure that no laser light reaches the detector. Figure 4.3 b) depicts the transmission characteristics of those filters and the beam splitter. The entire detection unit is based on the tube system from Qioptiq to avoid stray light in the detection path. Special support units were used to mount the filters so they can be easily exchanged. The filter, the focusing lens, and the pinhole are placed directly behind the folding mirror, as shown in figure 4.3 a). A spatial filter module with spring-loaded, backlash-free adjusters was utilized to provide precise xy-adjustment for the pinhole, as well as adjustment along the optical axis. The detection system was then coupled to the detector with a C-mount thread.

4.1 F-LSM Demonstration System Layout

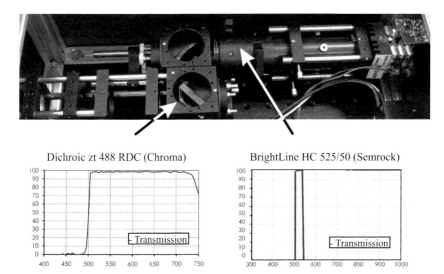

Fig. 4.3: a) Photograph of the detection unit and b) the transmission spectra of the dichroic and band-pass filters, respectively

A special holder was designed to bring the target slide with the fluorescent sample into the focal plane of the objective lens. This holder allows for axial and lateral adjustment and consists of three parts, the ground plate, the intermediate table holder and the slide holder, as depicted in figure 4.4. A 46 mm diameter hollow, cylindrical attachment with an M38 × 0.5 mm internal thread is mounted on the ground plate to which the intermediate sample holder is axially screwed in. In addition to connecting the ground plate with the intermediate sample holder this assembly's purpose is to axially position the microscope slide in the focal plane of the objective by screwing the intermediate sample holder in or out until the correct distance to the objective is attained. Since the resolution of the axial adjustment is limited by the thread pitch, a small one was chosen. Final positioning in axial direction is achieved with the piezoelectric actuator on which the objective is mounted. In addition to providing axial positioning, the intermediate table holder also provides a one-directional lateral adjustment of the slide holder via a guide rail in combination with screw threads for attaching it. This assembly permits the slide holder to be moved into the correct position and anchored. Adjustment in the remaining lateral direction of the table holder is achieved with an additional perpendicular guide rail for the target slide. The target slide is held with two steel spring clips, as shown by the photograph in figure 4.4 a). For the exhibitions or trade fairs, these spring clips are removed and replaced by a cover plate for purposes of laser safety. This

4 F-LSM System Realization

complete sample holder is mounted and adjusted with the help the tube system in front of the objective. After the objective and the z-shifter are removed, the tube system is

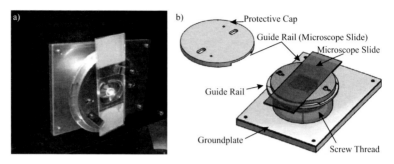

Fig. 4.4: a) Photograph of the target slide holder during illumination and b) the corresponding computer-aided design of the holder with its matching protection cover for trade fairs.

extended and the sample holder is placed so that the tube fits into the clear opening of the table holder. At 30 mm, this opening is the same size as the outer diameter of the tube system. Subsequently, the entire device is screwed on the front panel of the microscope, the additional tubes are removed and the objective and z-shifter are remounted. An in-house electronics module integrated in the microscope set-up itself is used for final image acquisition, as depicted in figure 4.5. This electronics module encloses the mirror start-up circuitry, mirror control and position sensing unit as well as the control unit for sampling and converting detector signals into pixels. The system platform is based on a field-programmable gate array (FPGA) for flexibility; an analog-to-digital converter is integrated. The start-up circuitry is necessary due to the special excitation behavior of the MEMS mirror [30]. Its particular properties require that the driving oscillation is first triggered at an initial frequency around 1.3 times the target frequency, close to the resonant frequency of the mirror, and then is gradually decreased. Once the mirror reaches the final frequency, it is continuously driven by rectangular voltage pulses of twice the scan frequency at every turn around point of the mirror plate. The mechanical oscillation and hence the geometrical position of the mirror is directly correlated to the applied driving impulses.

This dependence permits highly accurate computation of position, excepting for a discrete phase shift between the drive signal and the zero-crossing of the deflection angle. This phase shift is determined by the physical properties of each individual mirror and has to be adjusted as one of the system parameters prior to data acquisition. This initial calibration is typically be done using a sample with a known pattern in the image plane such as a lineal raster. However, with a 490 µm × 490 µm field-of-view, the size of the

4.1 F-LSM Demonstration System Layout

fluorescent structures for calibration should be correspondingly small. Such structures are difficult to position in front of the microscope objective.

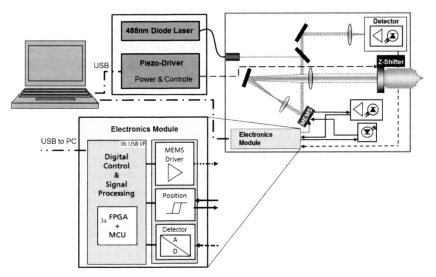

Fig. 4.5: Principal set-up: laptop, remotely connected power and control unit, and the microscope system with the integrated electronics module

For this reason, an alternative approach was chosen. Instead of placing the calibration structure in the image plane, a pinhole was placed in the intermediate focal plane of the telescope optics and an ordinary fluorescence reference slide was placed in the image plane instead, as shown in figure 4.6 a). The focal plane is easy to access and positioning the pinhole is very convenient. The bright central image point corresponding to the pinhole aperture splits into two or four points depending on whether the phase error is one or both scan directions, as shown in figure 4.6 c).

As a result, both phase errors can be distinguished simultaneously, and both of them can be corrected separately and instantaneously with direct feedback via the images. After calibration, the parameters are stored and the exact coordinates of the scanned pixels are computed. Due to the fact that the oscillation of the integrated mirror is extremely stable, no further calibration necessary and there is no immediate need for closed-loop operation in which the mirror position is continuously measured. However, if monitoring proper operation of the system in real time is required, it is possible to implement this. A projection module developed by IPMS uses a feedback signal from the mirror module back to the systems electronics [30].

4 F-LSM System Realization

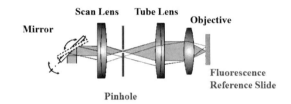

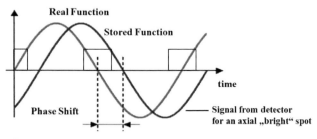

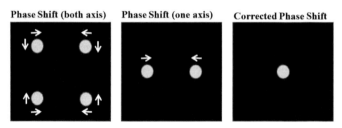

Fig. 4.6: a) Schematic optical train of the calibration set-up b) illustration of the effect when the real and stored mirror position functions differ and c) schematic illustration of this phase differences on the appearance of a pinhole image

During detection and post-processing, the image is formed one pixel at a time. Starting point is the analog electrical signal from the detector, which is a continuously varying voltage corresponding to the fluorescence intensity passed by the pinhole aperture. These incoming analog signals are transformed by the A/D converter into digital information and then periodically sampled and processed by an FPGA. The data processing unit implemented inside the FPGA transforms these digital signals of the special Lissajous pixel order of the scanning system into a conventional pixel order with eight-bit gray scale values.
According the maximum achievable optical resolution of 931 × 857 pixels discussed in

62

4.1 F-LSM Demonstration System Layout

chapter 3.1 an image format of 1024 × 1024 pixels is chosen with respect to the digital data processing within the FPGA. The complete image is stored and for any new measurement result for a pixel, the corresponding storage entry is overwritten. Instead of waiting until all pixels are overwritten, these images with partly overwritten pixels are repeatedly sent around twenty times per second via USB to the computer. The refresh rate for the complete 1024 × 1024 pixel frame is in contrast around 1 frame per second. This framerate is dependent on the defined pixel-array size as well as the filling factor of the Lissajous figure and the percentage of pixels which change and are renewed.

The transfered image can be stored, displayed, or alternatively superposed over a user-defined period using a Microsoft Windows XP application developed in-house. In addition to providing post-processing of the images, the purpose of this application is to set up and control the electronics module. Direct access to the electronics module from the PC is possible via this application by changing the USB port address definitions. The USB interface is connected to the FPGA and set-up parameters are stored in internal registers of the FPGA or image data are forwarded from two dual-port RAMs to the PC, based on the USB port address.

The entire electronics module is integrated in the F-LMS microscope set-up itself. The housing of the microscope is constructed with aluminium profiles in combination with aluminum side panels and a laser safety cover over the inspection port. A photograph of this system as displayed at a trade fair is shown below in figure 4.7.

Fig. 4.7: Photograph of the F-LSM demonstration system

4 F-LSM System Realization

4.2 Quantitative Optical Performance Testing

The imaging performance of the F-LSM system is first quantitatively tested in lateral direction with a standard United Stats Air Force (USAF) 1951 resolution target. Such a resolution target contains groupings of highly reflective vertical and horizontal chrome bars with known spatial frequencies on top of a 1 mm glass sample. With this traditional and simple method the resolution is easily tested by checking the highest resolvable spatial frequency. Due to the fact that this sample is non-fluorescent is was carefully placed on top of an green Fluor − Ref^{TM} fluorescent plastic slide (Plano GmbH, Wetzlar) with the chrome bars in close contact to it.

This fluorescent slide has an excitation peak and emission peak at 488 nm and 519 nm, respectively and is thus well suited. However, the thickness of the USAF resolution target exceeds that of a cover plate (170 µm) significantly. For this reason the microscope objective was first tested separately in a test set-up depicted in figure 4.9 to ensure the feasibility to bring the chrome structure in the focal plane of the objective and to image it. As shown in figure 4.9 a 1 mm thick test target with a crossing was placed on a dispersion plate in the focal plane of the objective and illuminated by a green LED light source. With the objective an image of this target is than formed and recorded with a CCD camera. In figure 4.9 b) and c) images of this structure are displayed for the two possible orientations - the crossing in contact to the dispersion plate and the reverse case.

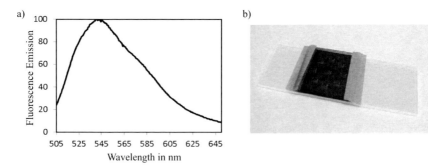

Fig. 4.8: With a Varian Cary spectrometer measured fluorescence spectrum of the Fluor-Ref fluorescent slide upon excitation with 488 nm (right) and photograph of the USAF resolution target mounted on top of the fluorescent slide(left)

64

4.2 Quantitative Optical Performance Testing

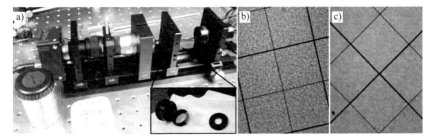

Fig. 4.9: a) Test set-up for the microscope objective to ensure the feasibility to bring the chrome structure in the focal plane of the objective, the images of a 1 mm test structure are displayed in b) and c) for the two possible orientations - the crossing in contact to b) the rough side and c) the polished side of the disperion plate

As seen in these images both times the crossing is located in the focal plane and imaged with high quality independently whether an effective 1 mm glass plate is in between or not. In consequence the method of placing the USAF resolution target on top of a fluorescent slide (see figure 4.10 d) and e)) was considered to be valuable configuration for resolution testing.

In figure 4.10 a) the image of the USAF resolution target obtained with the F-LSM demonstration set-up is shown. From this image with the high-lighted intensity distribution of group 7 it is shown that the F-LSM demonstration set-up can resolve the smallest element 6 in the 7th group which has a line spacing of 228 line pairs/mm corresponding to a line width of 2.2 µm. A closer look on the structures indicates that the resolution is not limited by the optics but by the sampling of the detector signals in one direction. Possible reasons for this sampling errors in the range of $1 - 2$ µm are assumptions in the calculation of the laser beam position used for assigning a measured intensity value to a certain pixel. The accuracy of this calculation depends on the degree of correspondence between the stored oscillation function of the mirror and the actual performance with which the laser beam position in the image field is calculated. Here an ideal sinusoidal movement of the mirror was assumed and used for the open loop position determination. To confirm this conclusion, that sampling errors caused these shifts in the pixel order and not the optical performance of the imaging system an additional performance evaluation is necessary and will be presented in the following sections. Never the less, this resolution still meats the target specification and it will be shown later on in this chapter that this resolution is sufficient to image biological samples.

Besides the USAF resolution target additional test structures where used to confirm this result and the resolution limit. In figure 4.10 b) a microring resonator is depicted. Such

4 F-LSM System Realization

a microring resonator consists of a waveguide and a ring waveguide in close proximity to it. Both structures have a width of 1.8 µm. Next to it in figure 4.10 b) a scaled reticle with 1 mm scale is depicted and reveals the field of view being 490 µm × 490 µm.

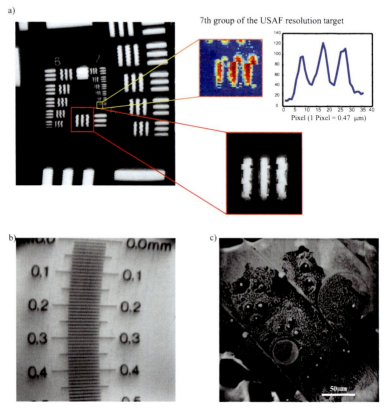

Fig. 4.10: Images obtained with the F-LSM demonstration set-up a) USAF resolution target b) scaled reticle with 1 mm scale and c) microring resonator stained with 1 µm diameter microspheres (F-13081; Molecular Probes)

4.2 Quantitative Optical Performance Testing

Axial Resolution Evaluation

Ring-stained microspheres pre-mounted on microscope slides were used as resolution targets (F-7237; Molecular Probes) for axial resolution testing. These fluorescent microspheres are specifically designed for examining the alignment, sensitivity, and stability of confocal laser-scanning microscopes and are particularly useful for testing the axial resolution of a confocal microscope [[88],[89]]. The reference standard used here consists of 15 μm diameter polystyrene beads with a well-defined fluorescing shell that is referred to as ring staining. When viewed with the F-LSM system, these beads consequently appear as a fluorescent ring of varying dimensions, depending on the focal plane. The change in ring diameter s with varying imaging depth is illustrated in figure 4.11a). The analytical relation between this ring diameter and the fluorescence bead radius and optical sectioning height h is given by:

$$s = 2 \times \sqrt{2 \cdot h \cdot r_{\text{bead}} - h^2} \qquad (4.2)$$
$$s = 11,19 \text{ μm} \qquad (4.3)$$

Thus, it is possible to assess the sectioning height by measuring the diameter s. The images of different sectioning heights taken with the F-LSM system are depicted in figure 4.12.

As illustrated in figure 4.13, the intensity distribution differ with the sectioning depths. Case A ($h \approx r$) is characterized by a narrow peak, whereas for B, the peak is broadened. This broadening can be explained by the finite axial resolution, which leads to the detection of photons below and above height h. This width of the intensity maximum was used for determining the axial resolution by calculating the sectioning height h for the three values of s given by the intensity distribution, maximum peak, and the two half-width.

Fluorescent Bead with Excitation and Emission Spectrum

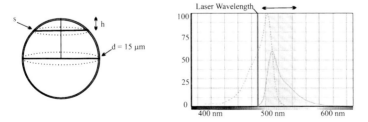

Fig. 4.11: a)Illustration of a fluorescent microsphere with the corresponding fluorescence spectrum (right)[65]

4 F-LSM System Realization

F-LSM Measurement

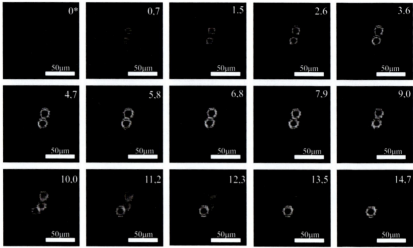

* Z-Shift Position in μm (from 0 to 14.7μm)

Fig. 4.12: Serial optical sectioning from top to bottom along the z-axis of ring-stained microspheres reveal the optical sectioning capabilities of the F-LSM demonstrator. The diameter of the fluorescent ring (or disc) seen is dependent on the depth of the optical focal plane.

Intensity Distribution for 2 Sectioning Heights

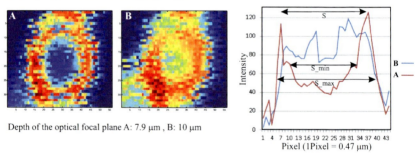

Fig. 4.13: Images of the upper fluorescent bead in two different sectioning depths with corresponding intensity distributions (right)

Note here, that only measurements with sectioning heights great enough to allow a measurement of all three diameters were used. Furthermore, images with sectioning

4.2 Quantitative Optical Performance Testing

heights h approaching the fluorescence microsphere diameter r were not considered. The reason for this is that for h approaching r, the uncertainty of the calculation of h with the measured value for s is very high. This is obvious when considering the small change in s for h approaching r. Figure 4.14b) shows calculation results with Mathematica for the minimum and maximum error. Here, the minimum and maximum error estimation was used for calculating the propagation of uncertainty instead of using the Taylor expansions. Reason is, that the function $h(r, s)$ is not differentiable for h approaching r.

$$\Delta h = Max\, |h(r', s') - h(r, s)|: \quad (4.4)$$
$$r - \Delta r \leq r' \leq r + \Delta r \quad (4.5)$$
$$s - \Delta s \leq s' \leq s + \Delta s \quad (4.6)$$

a) Δh Measurement and Stack of Intensity Profiles of Upper Bead along the Diameter of the Circle

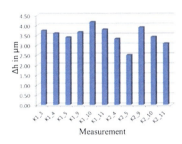
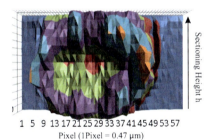

b) Minimum and Maximum Error for h

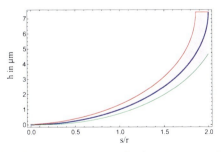

Fig. 4.14: a)Δh for eleven measurements (left) and a stack of intensity profiles of the upper bead in figure 4.12 along the diameter of the circle axial (right); b) measurement uncertainty calculation result with Mathematica for h with $\Delta s = 2$ μm and $\Delta r = \pm 0.07$ μm[65]

4 F-LSM System Realization

Tab. 4.2: Axial resolution measurements

Measurement	K1_3	K1_4	K1_5	K1_9	K1_10	K1_11	K2_4	K2_5	K2_9	K2_10	K2_11
S (µm)	9.84	11.72	13.13	13.13	11.72	9.38	11.25	13.13	13.13	12.19	10.78
S_min (µm)	3.75	4.69	11.25	10.78	4.22	3.28	6.09	11.25	10.31	9.38	7.03
S_max (µm)	13.13	13.13	14.53	14.53	13.59	13.13	13.13	14.06	14.53	14.06	13.13
h_min (µm)	0.24	0.38	2.59	2.33	0.31	0.18	0.66	2.59	2.09	1.67	0.89
h_max (µm)	3.98	3.98	6.00	6.00	4.47	3.98	3.98	5.09	6.00	5.09	3.98
Δh in µm	3.74	3.60	3.40	3.67	4.17	3.80	3.32	2.50	3.90	3.42	3.09

Mean height difference: 3.5 µm   **Standard deviation: 0.42 µm**

In table 4.2 the different measurement results for both fluorescence beads are summarized. With an average focal spot diameter of 3.5 µm and a standard deviation of 0.4 µm, this result proves that the axial performance of the system is within the target specifications. Considering measurement uncertainties, the result of an axial resolution of 1.7 µm is also close to the simulated axial resolution of $z_r = 1.2$ µm. In conclusion, these results corresponds well with the simulation results and demonstrate adequate resolution for cell imaging.

Another test sample in figure 4.16 shows a v-groove etched in silicon and filled with 1 µm fluorescent beads. Using ImageJ (open source) for post-processing, several raw image files uncorrected for distortion were combined and lead to a 3D-image of the v-groove as seen in figure 4.16 a) and b)[90]. However, the images also indicated that the fluorescent beads stuck together and filled the v-groove inhomogeneously. This effect first seen in the images of the F-LSM could be confirmed with images taken with the digital microscope VHX-2000 from Keyence (see figure 4.15).

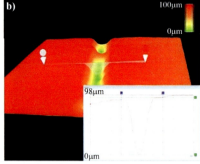

Fig. 4.15: Wide-field microscopy (Keyence, Digital Microscope VHX 2000) image of a v-groove filled with microspheres

4.2 Quantitative Optical Performance Testing

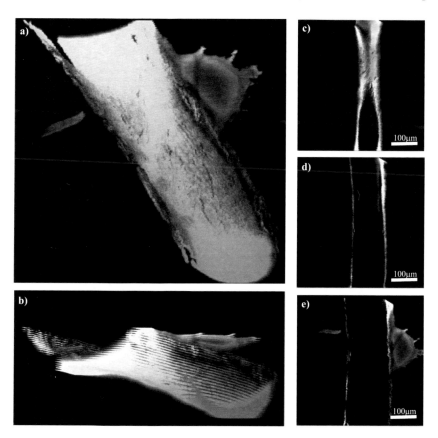

Fig. 4.16: Images of a V-groove etched in silicon and filled with 0.4 µm fluorescent microspheres a) and b) with the software ImageJ created stack of images that were taken with the F-LSM system in different focal depths [90] c) - e) images of the v-groove taken at different focal depths

In conclusion the performance analysis of the axial and lateral resolution of this F-LSM system gives functional verification of the MEMS-based approach and reveals an image resolution better than 2 µm in both directions and thus meets the target specifications defined in chapter 2.

4.3 Wavefront Error Measurement with a Shack-Hartmann Sensor

Due to the fact that the quantitative evaluation of the lateral resolution indicated, that the lateral resolution is probably limited by the laser beam position calculation and not by the imaging optics a second method for the performance evaluation is used. This additional performance evaluation is used to confirm the interpretation of the measurement results derived on the test samples.

Appropriate methods for the evaluation of the microscope optics are the application of interferometry or direct wavefront error measurements at the exit pupil using a Shack-Hartmann sensor. Considering the fact that the table supporting the microscope set-up is not isolated from vibration and that interferometry is highly vibration sensitive, a Shack-Hartmann sensor was chosen for this task. In addition to being relatively insensitive to vibration, a Shack-Hartmann sensor is a simple and compact direct wavefront measurement method by comparison.

The basic principle of a Shack-Hartmann sensor dates back to 1901 when the measurement technique was first introduced and proposed for optical system evaluations [91]. Early Hartmann geometry was based on an array of apertures in front of a detector. In 1971, it was modified by Roland Shack and Ben Platt by replacing this array with an array of small lenses called lenslets, resulting in the Shack-Hartmann sensor today [[92],[93]]. The principle of measuring the wavefront slope with the Shack-Hartmann sensor is depicted in figure 4.17.

An incident wavefront is transformed into a grid of focal spots on the detector array by the lenslets, which is generally a CCD detector. The position of these focal spots is directly related to the average gradient of the wavefront at each single lens. The incident wavefront slope is determined by measuring the focal spot positions.

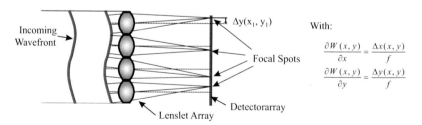

Fig. 4.17: Basic layout of a Shack-Hartmann wavefront sensor

4.3 Wavefront Error Measurement with a Shack-Hartmann Sensor

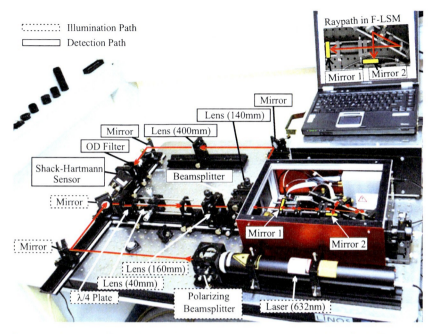

Fig. 4.18: Photograph of the wavefront measurement set-up with the Shack-Hartmann sensor consisting of the illumination path, the F-LSM microscope system with Mirror 1 and Mirror 2 for the reference and wavefront measurements respectively, and the detection unit with the Shack-Hartmann sensor

As introduces in chapter 1 the integration of the wavefront slopes produces wavefront optical path difference (OPD) maps in the same way the derivative of the wavefront (OPD) map produces wavefront slopes. The measured wavefront can then be compared with the wavefront predicted by simulations in chapter 3.2. As a result, the basic analysis process consists of three steps: spot position determination, conversion to the wavefront slopes, and finally reconstruction of the wavefront and OPD.

The optical set-up to perform this measurement is depicted in figure 4.18. It consists of three parts:

1. the external illumination path to couple the light of a HeNe-laser into the microscope system,
2. the microscope system itself with the objective removed, and
3. the detection optics, which image the light emerging from the exit pupil of the microscope onto the Shack-Hartmann sensor (SHSLab, Optocraft).

73

4 F-LSM System Realization

To ensure separation of the illumination and the detection optics, a beam splitter is placed at the target end of the microscope where the HeNe-laser beam enters the detector. For comparing the measured wavefront with the wavefront predicted by simulations, it is important that the beam diameters are the same size at the exit pupil of the F-LSM system. This is the reason a telescope optics consisting of two lenses with 40 and 160 mm focal lengths are used for illumination and magnification of the laser beam diameter. With a beam diameter of around 1.2 mm this leads to an diameter of 8.4 mm in the exit pupil of the microscope. In the demonstration system this diameter was chosen to be smaller than 7.7 mm due to the fact that a larger beam diameters result in diffraction effects at the mirror site as discussed in chapter 3.1. The resulting effects in the wavefront measurement would need to be subtracted before an analysis and comparison of the microscopes imaging optics performance with the simulation results is possible. This is hard to achieve.

Beside the magnification optics, an optical isolator is included in the illumination path to avoid reflection of the laser light at Mirrors 1 and 2 back into the laser itself. The optical isolator consists of a linear polarizer in combination with a $\frac{\lambda}{4}$ plate whose axis is oriented at 45° to the polarization direction of the laser beam passing through the polarizer. This leads to a circularly polarized laser beam after the $\frac{\lambda}{4}$ plate which will change its orientation when reflected back on Mirrors 1 or 2. As a result, the light reflected back will be linearly polarized with a shift of 90° from the original orientation when it passed the $\frac{\lambda}{4}$ plate. It will therefore be blocked by the polarizer in front of the laser.

To ensure that the incident illumination light on Mirror 2 and the reflected light pass through all microscope optics centrally with normal incidence along the optical axis, highly accurate adjustment of the illumination optics is necessary.

For the final measurement, the two Mirrors 1 and 2 are placed into the optical train of the F-LSM. First, Mirror 1 is placed at the exit pupil of the microscope for the reference measurement. In this reference measurement, the HeNe-laser light is directed onto Mirror 1 and is then deflected onto the Shack-Hartmann sensor. In consequence, this measurement result corresponds to wavefront aberrations caused by the optics in the illumination and detection paths.

Afterwards, the final measurement of the wave-front aberrations caused by the F-LSM optics is performed by placing Mirror 2 in front of the dichroic beam splitter inside the microscope and subtracting the reference measurement. In consequence the resulting $PV_{\mathrm{microscope}}$ values depicted in figure 4.19 d) correspond to the wavefront aberrations caused by the F-LMS optics. For each reference measurements, three final measurements with Mirror 2 were performed and the arithmetical mean was taken. In total this complete measurement, with first taking the reference measurement and then performing the final measurement from which the reference measurement is substracted,

4.3 Wavefront Error Measurement with a Shack-Hartmann Sensor

a) Shack-Hartmann Wavefront Measurement Result in Terms of the Wavelength λ

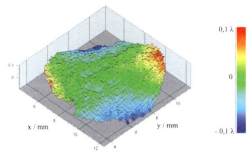

b) Typical Wavefront Deviation Described with Zernike Polynomial Coefficients

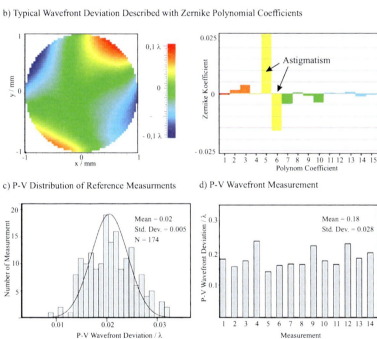

c) P-V Distribution of Reference Measurments   d) P-V Wavefront Measurement

Fig. 4.19: a) Shack-Hartmann wavefront measurement result in terms of the wavelength λ, b) reconstruction of the wavefront measurements from a) above using Zernike polynomials on the left and the corresponding Zernike coefficients on the right c) P-V wavefront measurement variations from 174 reference measurements, and d) the P-V wavefront deviations from the fourteen final measurements

4 F-LSM System Realization

was performed fourteen times. In figure 4.19 a) a typical measured wavefront in the exit pupil of the F-LSM system is depicted and in d) the corresponding P-V wavefront deviation values of the fourteen final measurements are depicted. The resulting mean PV_{F-LSM} value of these measurements was 0.18 wavelengths with a standard deviation of 0.028 wavelengths.

To assess the repeatability of a measurement, a reference measurement with Mirror 1 was taken and instead of removing Mirror 1 for the final measurement, it was left in place and the deviations in the PV values from this were compared to 174 subsequent measurements using the same set-up. The resulting P-V wavefront deviations from this initial reference measurement are depicted in figure 4.19 c) and exhibit a typical normal distribution with an average PV wavefront deviation of 0,020 wavelengths.

The same kind of measurement was done with Mirror 2 placed in front of the dichroic mirror, leading to similar measurement uncertainties of an average PV wavefront deviation in the range of 0,021 wavelengths, corresponding to around 10 nm. This finite precision comes about as a result of finite pixelization, detector signal-to-noise ratio, CCD-readout noise, background light, vibrations, laser fluctuations and several other effects [94].

Taking the propagation of uncertainties into consideration, the standard deviation of the measured difference $PV_{microscope} = PV_{final} - PV_{reference}$ in 4.19 d) is expected to be

$$\Delta PV_{F-LSM} = \sqrt{\Delta PV_{final}^2 + \Delta PV_{reference}^2} = 0,029 \qquad (4.7)$$

Hence, the standard deviation of 0,028 wavelengths in the 14 individual measurement results shown in figure 4.19 d) are within the calculated uncertainty.

To compare these measured results with the simulations, the wavefront OPD was reconstructed using Zernike polynomials. A sample reconstruction corresponding to one of the fourteen measurements is shown in figure 4.19 b). These Zernike polynomials present an alternative description of the classical wavefront aberrations inherent in an optical system and are often used for data fitting. It is advantageous to use them instead of the historical power series expansion from Seidel due to the fact that the Zernike polynomials form a complete set of orthogonal basis functions.

However, care has to be taken with regard to which set of the Zernike polynomials is used for data fitting. Even though there is a DIN ISO specification, there are currently at least four other Zernike polynomial sets in use. In the case here, the notation defined in R. Noll, 'Zernike polynomials and atmospheric turbulence' [83] was applied to ensure the comparability with the simulation results. These Zernike polynomials are orthonormal, with the first terms being related to the traditional Seidel aberrations as listed in table 4.3 [[80], [83]]. Further information about their definitions can be found in [80].

It can easily be seen in the wavefront deviation as well as the reconstruction with Zernike polynomials figure 4.19 b) that astigmatism is the dominant aberration in this system.

Tab. 4.3: Seidel aberrations corresponding to the first Zernike polynomials [[83], [80]]

Order	Zernike Polynomials	Corresponding Aberrations
Z_1	1	piston
Z_2	$2\rho \cos \Theta$	x-tilt
Z_3	$2\rho \sin \Theta$	y-tilt
Z_4	$\sqrt{3}(2\rho^2 - 1)$	defocus
Z_5	$\sqrt{6}\rho^2 \sin 2\Theta$	astigmatism at 0° & focus
Z_6	$\sqrt{6}\rho^2 \cos 2\Theta$	astigmatism at 45° & focus
Z_7	$\sqrt{8}(3\rho^3 - 2\rho) \sin \Theta$	coma & x-tilt
Z_8	$\sqrt{8}(3\rho^3 - 2\rho) \cos \Theta$	coma & y-tilt
Z_{11}	$\sqrt{5}(6\rho^4 - 6\rho^2 + 1)$	3rd-order spherical & focus

The presence of both 45° and 0° astigmatism is explained by the placement of the Shack-Hartmann sensor which could not be mounted parallel to the optical table due to very limited space. Reason for the strong degree of astigmatism as the dominant wavefront aberration can be mirror deformations as discussed in chater 3. In contrast, no reasonable displacement of the lenses can cause this strong on-axis astigmatism. Due to the fact that the integrated mirror is from an old batch (2003) it was not tested beforehand. For this reason 12 mirrors from the same batch of different wafers were tested in the following using interferometric measurements (NT1100, Veeco).

Tab. 4.4: MEMS mirror curvature: interferometric measurement results

Mirror	Curvature x-axis	Curvature y-axis	P-V value
SC07	-1.33 m	$-1,32$ m	379 nm
SC18	-2.42 m	-2.47 m	202 nm
SC24	-2.26 m	-2.34 m	214 nm
SC19	-3.29 m	-3.07 m	163 nm
SC30	-2.58 m	-2.43 m	206 nm
SC27	-1.56 m	-1.50 m	333 nm
SC32	-1.45 m	-1.40 m	357 nm
SC04	-1.84 m	-1.88 m	266 nm
SC13	-2.43 m	-2.41 m	207 nm
SC10	-1.50 m	-1.47 m	340 nm
SC21	-1.55 m	-1.54 m	325 nm

4 F-LSM System Realization

a) Interferometric Image of a Dual-Axis MEMS-Mirror

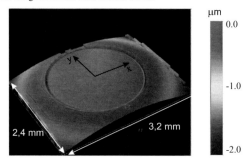

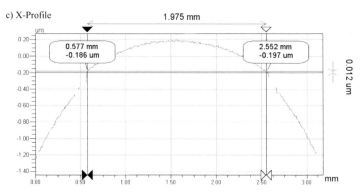

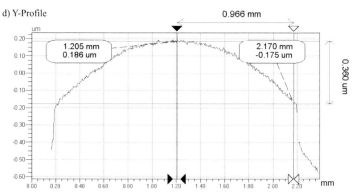

Fig. 4.20: a) Interferometric measurement and reconstruction of a dual-axis MEMS mirror and measurement results of the x and y profiles in b) and c) revealing mirror curvature of 1.5 m

4.3 Wavefront Error Measurement with a Shack-Hartmann Sensor

It turned out that these mirrors indeed exhibit different degrees of distortion, with radius of curvatures ranging from 1.3 m to 3.3 m. These measurement results, including the radius of curvature and the P-V values, are listed in table 4.4. In figure 4.20, a 3D reconstruction of a typical interferometric measurement from one of the mirrors is depicted.

Assuming a mirror curvature of 2 m in the simulation and taking a moderate decentration of 200 μm of the same into account, the simulated wavefront measurement fits very well with the Shack-Hartmann measurements. Figure 4.21 gives a comparison of both the simulation and the measurement result. Even though the mirror exhibits some curvature, OPD wavefront deviations of less than 0.2 wavelengths still indicate a diffraction limited performance for a field angle of 0°. In the future, mirrors from more recent batches should be used and tested before integration.

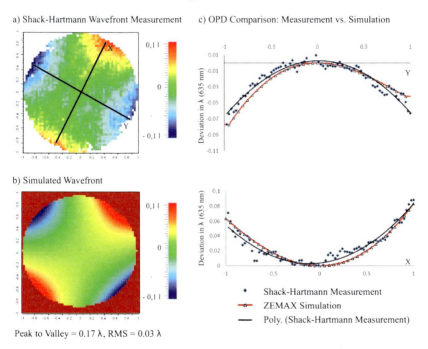

Fig. 4.21: Comparison of the Shack-Hartmann measurement result a) with the simulation results b). The corresponding OPD graphs are shown in c).

4 F-LSM System Realization

4.4 Images of Biological Samples

In addition to the images taken for qualitative evaluation of the F-LSM system performance, this system was used for several biological imaging experiments. The images presented in the following figures further demonstrate the earlier results with respect to the depth discrimination and the resolution capability of the F-LSM demonstration system. In figures 4.23 a) - f), images from a stained histologic skin sample are depicted, kindly provided by the Pathology Department of the Friedrichstadt Hospital in Dresden. This 15 µm thick histological section of human skin is H&E (hematoxylin and eosin) stained and sandwiched between a microscope slide and coverslip [95]. A special characteristic of this common laboratory stain is the fluorescence when illuminated with 488 nm light. Reason for this fluorescence appearance is eosin, which itself is a fluorescent dye [96]. To measure the corresponding fluorescence emission spectrum depicted in figure 4.23 when illuminated with 488 nm a Varian Cary spectrometer was used. Figure 4.22 shows the resulting fluorescence spectrum and a wide-field microscopy image of the sample. An image taken with the F-LSM demonstration system from an area of this histological section is depicted in figure 4.23 a). The different layers of the epidermis are distinguishable and labeled. It should be noted here that the several epidermal layers are not distinctly different tissues but rather reflect visible stages of differentiation along the outward displacement of the cells through the epidermal layers. The dermis, which consists of connective tissue and cushions the body from stress and strain, is the underlying thicker layer and has a different origin and is thus different tissue. In addition, various appendages including sweat glands, sebaceous glands, hair and hair follicles are seen in this histological section shown in figures 4.23 b) - f).

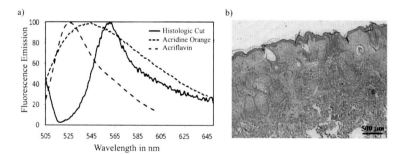

Fig. 4.22: (a)Fluorescence spectrum of the stained histological skin sample in comparison to acridine orange and acriflavin stained samples (illumination with 488 nm);(b) wide-field microscopy image of the sample (Nikon Eclipse L300D)

4.4 Images of Biological Samples

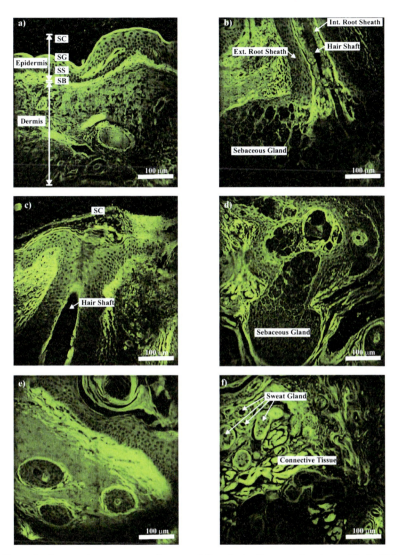

Fig. 4.23: Images from sections of a stained histologic skin sample taken with the F-LSM demonstration set-up; (a) section of the epidermis and the dermis with the different layers labeled; (b) longitudinally cut profile of a hair follicle with sebaceous glands; (c)-(e) different appendages of the skin (f) cut profile of sweat glands embedded in the connective tissue of the dermis

4 F-LSM System Realization

As can be seen in the surrounding tissue structure, these appendages of the skin are epidermal invaginations into underlying skin layers. In figure 4.23 b), a longitudinal section of a hair follicle with the central lumen containing the hair shaft, the internal and external root sheath, and the associated sebaceous glands opening into the lumen are depicted. These sebaceous glands are outgrowths of the external root sheath and secrete an oily substance into the follicular space. The above-mentioned sweat glands embedded in the connective tissue of the dermis are depicted in figure 4.23 f).

In addition to this histological section, images were taken of "thick" cross-sectional samples from different plants, such as the stem of convalaria (lily of the valley) and the flowers of lavender and tensile, as depicted in figure 4.25. All these samples where provided by Mr. Dethloff, who manually cut, stained, washed the samples with distilled water, and fixed them afterwards with superglue on a microscope slide. Acridin orange was used for staining convalaria and a stain known as W-3A developed by Mr. Wacker at the University of Wurzburg was used for the lavender and tensile flowers. Both are typical stains used in microscopy, but not generally for fluorescence microscopy. However, the fluorescence emission of acridin orange is well known and is used for differential staining of DNA and RNA in intact cells and isolated cell nuclei. Due to its emission maximum at 530 nm when excited with blue light, acridin orange was especially well-suited for imaging with the F-LSM. Figure 4.24 a) - c) shows a row of images from the acridin-orange stained convalaria stem. These images further demonstrate the depth discrimination as well as the changes in the cellular structures vs. depth. For demonstration purposes, one cell is stained red to make visible the structural change in these three images, taken with depth differences of 3 μm in each case. Figure 4.25 a), e)-f) shows images from lavender and tensile flowers. The W-3A staining method was used for these samples; three dyes, acriflavin, acridin red and astra blue are applied to label specific histological features of plant tissue.

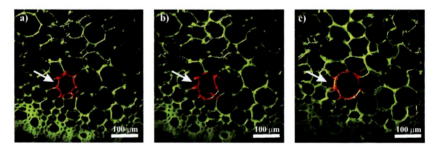

Fig. 4.24: Series of images taken with the F-LSM demonstration system of an acridin orange stained convalaria stem (depths difference between the different images: 3 μm)

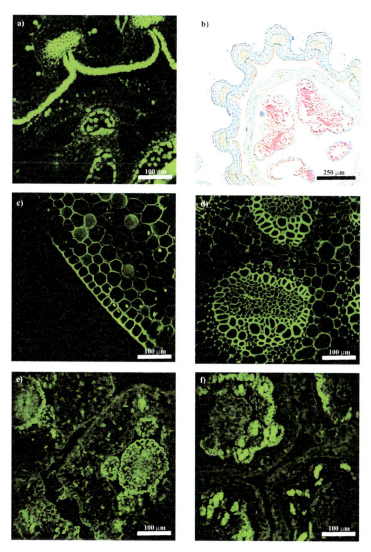

Fig. 4.25: Images of "thick" cross-sectional samples from different plants; a)F-LSM image of a tensile flower (b)wide-microscopy image of tensile c)-d) F-LSM images of different sections from a convalaria sample; e)-f) cross-sectional F-LSM images of lavender

4 F-LSM System Realization

This advantageous approach of selective staining in wide-field microscopy is not as effective for F-LSM imaging, since not all of these dyes fluoresce in the range of 500 − 550 nm. A comparison of the wide-field microscopy image for the tensile flower depicted in figure 4.25 b) with the F-LSM image in figure 4.25 a) reveals that only the light pink-orange structures fluoresce. These structures are stained with acriflavin, whereas the deep blue-green and brilliant red parts are stained with astra blue and acridin red, respectively. Even though the emission maximum of acriflavin occurs at a longer wavelength than acridin orange, it still shows fluorescence in the range of 505 − 545 nm and is thus visible in the F-LSM images (see emission spectra in figure 4.22). Acridin red, in contrast, fluoresces as well but is not visible due to its emission maximum at much longer wavelengths (> 600 nm). All of these images demonstrate good sectioning capability of the F-LSM demonstration system for biological cellular structures.

4.5 Summary

In summary, the F-LSM demonstration set-up shown in figure 4.26 has fulfilled the design requirements listed in chapter 1. Using a USAF 1951 test chart and stained microspheres, quantitative measurements of the axial and lateral imaging performance where carried out and revealed a resolution of around 2 µm and 1.7 µm in lateral and axial direction, respectively. Wavefront measurements with a Shack-Hartmann sensor confirmed that the lateral resolution is not limited by the optical performance but instead by the calculation of the laser beam position used for assigning a measured intensity value to a certain pixel. The accuracy of this calculation depends on the degree of correspondence between the stored oscillation function of the mirror and the actual performance with which the laser beam position in the image field is calculated. The wavefront measurements could also prove a diffraction limited on-axis performance of the F-LSM system. Additional images made from a histological section and biological samples further demonstrated the F-LSM system capability and performance. The final specifications are listed in Table 4.5. The assembly of the F-LSM demonstration set-up in a small package (26 cm × 15 cm × 34 cm (w x h x d)) was challenging and necessitated several custom adapters and a design that allowed lateral and axial adjustment for the target holder. Total fabrication and assembly costs of the system shown in figure 4.26 are less than 40,000 EUR. This figure does not include the electronics, nor the design and testing. For the application in a clinical setting a hand held solution is necessary. For this reason the second objective of this PhD research project is to explore how the optical system design of the first F-LSM system can be optimized to achieve a handheld F-LSM system with an extended field of view. To this end, a second design with a customized scan lens design is presented in the following. This system is handheld and offers a more than two times larger field-of-view of 750 µm × 750 µm.

4.5 Summary

Tab. 4.5: F-LSM Demonstration System Specifications

Parameter	Symbol	Value	Unit
$\lambda_{\text{Illumination}}$	$\lambda_{\text{Illum.}}$	488	nm
$\lambda_{\text{Fluorescence}}$	λ_{f}	520	nm
Mirror Diameter	D_{mirror}	2	mm
Lateral Resolution	$\Delta x, \Delta y$	2	µm
Axial Resolution	Δz	1.7	µm
Field of View		490×490	µm^2
Working Distance	d	1	mm
Axial Shift	h	100	µm
Size		$26 \times 14.5 \times 34$	cm^3

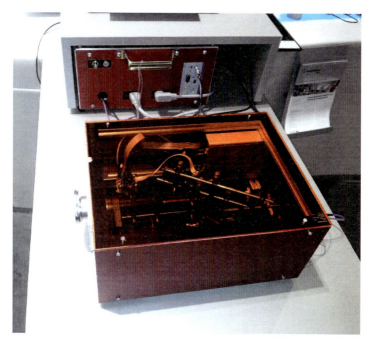

Fig. 4.26: Photograph of the F-LSM demonstration system on a trade fair

5 Transfer into a Handheld Microscope Design

The objective of the present chapter is to explore how the optical system design of the first F-LSM system, which provided the functional verification of the MEMS-based approach, has to be changed to achieve a handheld F-LSM system with an extended field of view. The design requirements for this are discussed and an approach for realizing a new optical system design is explored. It will be shown that an optical imaging system with custom-designed scan and tube lenses in combination with an alternative objective enable the field of view to be enlarged by more than twofold while achieving a significant reduction of the overall system size. A solution for a handheld MEMS-based F-LSM system will be presented in a theoretical system layout including the new imaging optics. The manufacturability of this design is confirmed via tolerance analysis. In a final comparison of the demonstration system presented in the previous chapter with this theoretical solution for a handheld system, the feasibility of the MEMS-based approach in combination with a custom-designed imaging system are outlined.

5.1 Optimized Imaging Optics

The objective here is to explore how to transform the F-LSM demonstration system presented in the last chapter into a handheld F-LSM solution with an enlarged field of view. The field of view of the demonstration system was limited by the optical elements to effective mirror scan angles of $\pm 5°$ and not by the maximum achievable scan angles of around $\pm 7.5°$. The main reason for the limited field of view of 490 µm × 490 µm was the optical performance of the relay optics, which consists of two off-the-shelf lenses as discussed in chapter four.

It is necessary for this reason to investigate modified imaging optics in order to achieve an extended field of view and to minimize the overall system layout. As described in chapter three and shown in figure 5.1, the imaging optics consist of a scan lens, a tube lens and the objective.

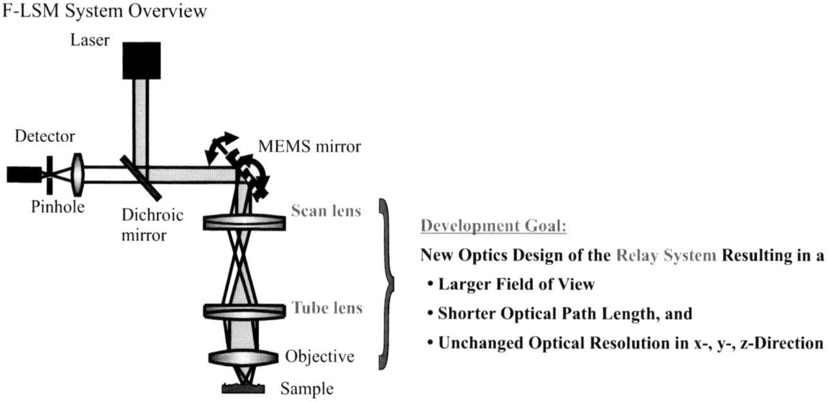

5 Transfer into a Handheld Microscope Design

F-LSM System Overview

Fig. 5.1: Schematic of optical train of the F-LSM system based on a dual-axis MEMS mirror where optical elements of the imaging system are highlighted

Development Goal:
New Optics Design of the Relay System Resulting in a
• Larger Field of View
• Shorter Optical Path Length, and
• Unchanged Optical Resolution in x-, y-, z-Direction

A two step approach for the modified optics design is adopted based on the correlations between these three different optical elements investigated in chapter three. First, the Hastings Triplet is removed and a new scan lens of equal focal lengths is designed to achieve diffraction-limited performance for mechanical MEMS mirror scan angles ϕ up to ±7.5°, while all other optical components remain unchanged. Under these conditions, the scan angles refer to the extended field of view of 750 µm × 750 µm originating from the relation for the length of one side:

$$d_{\text{im}} = 2 \cdot f_{\text{Objective}} \cdot \tan(\frac{2 \cdot \phi}{M}) \tag{5.1}$$

as introduced in chapter three, where M denotes the beam expansion ratio intermediate optics and $f_{\text{Objective}}$, the objective focal length. The possible design approaches and glass selection options for achieving an appropriate new scan lens design that permits this large field of view are evaluated first. The design and simulations are then performed within pre-defined boundary conditions using optical simulation software.

In the second step, the objective is replaced and a new tube lens is designed to reduce the overall optical path lengths and thus to minimize the total optical system size.

For a given target numerical aperture, the objective parameters define the necessary magnification factor of the relay system, and thus the tube lens focal lengths for a given scan lens. Replacing the CFI Plan Apochromat VC 20x 0.75 (Nikon, Japan) with a CFI S Plan Fluor ELWD 40x 0.6 (Nikon, Japan) with a focal length f_{Obj40} of 5 mm results

5.1 Optimized Imaging Optics

in a reduced laser beam diameter

$$D_{\text{Obj40}} = 2NA \cdot f_{\text{Objective}} = 3.85 \text{ mm} \qquad (5.2)$$

required at the entrance pupil of the objective for a numerical aperture of 0.385 as used for the demonstration system in chapter three. As a consequence, the resulting focal length of the tube lens is given by:

$$f_{\text{tube}} = \frac{D_{\text{Obj40}}}{D_{\text{laser}}} \cdot f_{\text{scan}} = 70 \text{ mm} \qquad (5.3)$$

which is half the focal length of the original tube lens in the F-LSM demonstration system and thus leads to a significant reduction of the overall optical path length. To achieve diffraction-limited performance over the extended field of view a custom-designed solution is necessary for the tube lens. The main challenges encountered in the design of a new scan and tube lens for the MEMS-based fluorescence laser scanning confocal microscope are:

1. achieving diffraction-limited, color-corrected performance for larger mirror scan angels of at least $\pm 7.5°$ in both directions while offering an optical resolution comparable to the one of the F-LSM demonstration system
2. realizing a scan lens design with a remote entrance pupil in combination with a small focal length
3. using of a small number of lens elements for both lenses to keep manufacturing cost reasonable

5.1.1 Scan Lens Design

For the scan lens design, the first two of the three requirements stated above are of special interest and difficult to achieve. Providing an external aperture while maintaining a high degree of aberration correction over a large field of view for both wavelengths, the fluorescence excitation wavelength and the fluorescence emission wavelength, is challenging.

The reason for this is that, due to the position of the external stop, odd wavefront aberrations such as coma distortion and lateral chromatic aberration are more difficult to correct compared to other imaging systems in which symmetry relations are used. To summarize, the requirements for the scan lens resemble them for an eyepiece to a high degree.

Eyepieces have to cover a fairly wide field of view through a relatively small aperture at a distance of 10 to 20 mm away from the final glass surface. The reason for the external aperture is that eyepieces need to be matched to the pupil of the human iris to avoid

5 Transfer into a Handheld Microscope Design

vignette effects while providing adequate clearance for the eyelashes of the observer. Eyepiece designs can be classified by number of elements, the apparent field of view and the eye relief. Two-lens eyepieces are of special interest for the scan lens design, since more complex solutions with more elements are accompanied by added cost and size. Representative two-lens designs are the Huyghenian, Ramsden, Kellner, König, Ploessl and Abbe Orthoscopic eyepieces [97].

Different 3- and 4-Element Eyepiece Designs

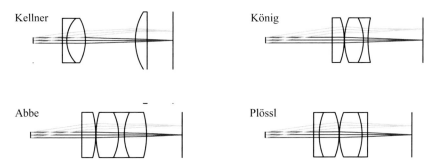

Fig. 5.2: Basic forms of three- and four-element optical eyepieces (data from [98])

Taking the required field of view and the color correction for both wavelengths into account, the Kellner, Ploessl, Koenig and Abbe Orthoscopic eyepiece come closer to the requirements. While the Kellner and Koenig eyepieces comprise a doublet in combination with a single lens, the Ploessl eyepiece consists of two doublets, while the orthoscopic eyepiece consists of a triplet in combination with a single lens in front as depicted in figure 5.2. The additional element in the Ploessl and Abbe Orthoscopic eyepieces creates a further degree of freedom for correction of aberrations and allows larger field of view as a result. In the end, the Ploessl design was chosen as a starting system.

The scan lens was then optimized as an element of the optical imaging system together with the tube lens, which was kept the same as in the demonstration system, and the objective was simulated as ideal lens. The optimization was carried out for both the fluorescence excitation and emission wavelengths simultaneously in multi-configuration mode of the software in order to account for the different mirror scan angles. A focal lengths with a fixed value of 20 mm was defined in the merit function, while surface curvatures and element thicknesses were varied over a controlled range.

For glass selection, a glass substitution template was used during the optimization process to exchange candidate glasses that met the requirements given by the template. Besides the primary interest of selecting a particular type of glass that minimize the remaining optical aberrations in the optical system and possesses maximum transmission

5.1 Optimized Imaging Optics

over the spectral region of interest, other important factors for the final manufacturing process are pricing and availability. To account for these, the choice of glasses was restricted in the substitution template to moderately priced glasses from well known vendors Schott AG and Ohara Group. Additionally glasses containing lead where excluded due to the fact that glass manufacturers are replacing them in their product ranges with lead-free glasses.

Another consideration in lens design for fluorescence applications is the inherent fluorescence of certain optical glasses. This fluorescence is generated by point defects like color centers, which are known to originate from rare earth elements, and certain impurities [99][100]. In order to compare the fluorescence of different optical glasses, the emission spectra are measured, integrated, and compared with those of a reference glass such as SF1 or SF6 [99]. The excitation wavelength for these measurements is generally 365 nm and typical emission maxima can be seen at e.g. 435 nm and 525 nm[99]. In cases where the excitation wavelength differs from the above, the results can only be seen as an indication of fluorescence within these glasses. Indeed, a further study by Schott AG has shown that the absolute level of fluorescence light is significantly decreased for higher excitation wavelengths such as 488 nm and 532 nm in most glasses. Thus, most glasses are not problematic except for N-SF57HHT, SF2, and SF6, which were excluded from the glass substitution template as a consequence.

The glasses selected for the scan lens design are listed next to the scan lens design in figure 5.3 b). This glass choice in combination with the lens shape was discussed with the lens supplier prior to the decision. In addition, the different radii of curvature for the single lens elements were matched to the manufacturing tools offered by the supplier to avoid additional tooling charges. The final lens design and different performance parameters are depicted in figure 5.3.

5 Transfer into a Handheld Microscope Design

a) Optical layout including the new designed scan lens

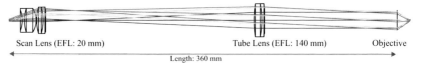

Scan Lens (EFL: 20 mm)   Tube Lens (EFL: 140 mm)   Objective

Length: 360 mm

b) Scan Lens Design Data

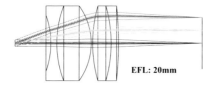

EFL: 20mm

	Surface	Radius	Glass
	1	-21.4	N-SF57
Element 1	2	27.7	N-LAK9
	3	-16.5	
	1	35.7	N-LAF34
Element 2	2	-35.7	N-BALF4
	3	-92.1	

c) Strehl Ratio of the original F-LSM Demonstration Systems in Comparison to the New Design

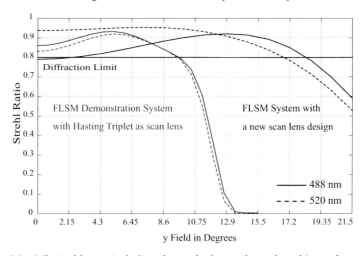

Fig. 5.3: a) Optical layout including the newly designed scan lens; b) scan lens specifications; c) Strehl ratio of the original F-LSM demonstration system in comparison to the new design

As introduced in chapter three, diffraction-limited performance in accordance with the Rayleigh criterion is achieved for a Strehl ratio larger than 0.8, corresponding to a peak-to-valley value for the wave aberration less than $\lambda/4$ [69]. As shown in figure 5.3 b), near diffraction-limited performance is present over the full field of view. In figure 5.4, the

5.1 Optimized Imaging Optics

optical path difference (OPD) is plotted for the different field angles and displays the different kind of aberrations present in the system more completely.

Optical Path Difference

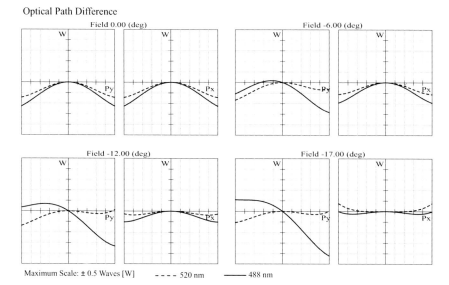

Fig. 5.4: Wavefront aberrations at the exit pupil of the objective for four different field angles

On-axis, spherical aberration is present with a certain amount of defocus from marginal focus. For large field angles, lateral chromatic aberration at 488 nm becomes the dominant wavefront aberration in this system. Lateral chromatic aberration is caused by dispersion of the chief ray and is sometimes referred as chromatic tilt. This term means that if the wavefront for 488 nm is regarded as the reference sphere, the wavefront for 520 nm is a sphere of the same radius, but tilted with respect to the latter.

Besides lateral chromatic aberration, coma and some field curvature are present as residual aberrations.

5.1.2 Tube Lens Design

In the next step of the design process, the 20x CFI Plan Apochromat VC objective (Nikon, Japan) is replaced with a 40x CFI S Plan Fluor ELWD objective (Nikon, Japan), and a new tube lens is designed to account for this change to achieve a larger field of view with diffraction-limited performance. The purpose for the integration of this new objective is to reduce the focal length by a factor two to 5 mm. This new focal lengths results in a reduced necessary magnification M=3.5 of the relay optics while achieving the same image space NA of 0.385 present in the demonstration system. As a result, the focal length of the tube lens is also reduced by the same factor, from 140 mm to 70 mm. A Hastings Triplet is used for better aberration correction at this reduced focal lengths. The objective was simulated as an ideal lens during the design process and the same glass template was applied as for the scan lens design. Lens thicknesses and radii of curvature were varied over pre-defined ranges. Later in the design process, all scan lens parameters were allowed to vary and the whole optical imaging system, including the tube and scan lens, was optimized to achieve the best overall performance for the imaging system. Finally, the radii of curvature for the lens elements in the optimized optical system were matched with respect to the tools provided by the supplier. The final specifications of the system are listed in table 5.1.

Tab. 5.1: Optical System Specifications

Type	Radius (mm)	Thickness (mm)	Glass	Semi-Diameter
Eye Relief		9.87		0.6
Scan Lens (Doublet 1)	-21.44	1.66	N-SF57	7.8
	27.68	9.24	N-LAK9	9.0
	-16.55			9.0
Air		0.40		8.3
Scan Lens (Doublet 2)	35.23	5.00	N-LAF34	10.0
	-35.74	2.00	N-BALF4	10.0
	-92.06			10.0
Air		0.00		8.8
		22.64		8.8
		66.74		7.4
Tube Lens	54.45	4.00	N-BAF4	10.0
	30.07	7.20	N-SK5	10.0
	-24.41	3.70	N-BAF51	10.0
	-97.51			10.0
Air		51.45		7.9
Objective		0.00		3.3
Image Distance		5.00		3.3

5.1 Optimized Imaging Optics

a) Optical layout of the demonstration system with catalog lenses

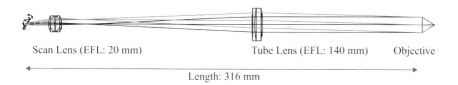

Scan Lens (EFL: 20 mm)   Tube Lens (EFL: 140 mm)   Objective

←——————————— Length: 316 mm ———————————→

b) Optical layout including the new custom-designed scan lens

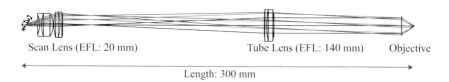

Scan Lens (EFL: 20 mm)   Tube Lens (EFL: 140 mm)   Objective

←——————————— Length: 300 mm ———————————→

c) New optical layout with a custom-designed scan and tube lens and a 40x Objective (Nikon)

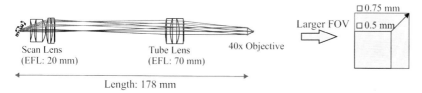

Scan Lens   Tube Lens   40x Objective   Larger FOV
(EFL: 20 mm)   (EFL: 70 mm)

□ 0.75 mm
□ 0.5 mm

←——————— Length: 178 mm ———————→

Fig. 5.5: Illustration of the different optical layouts presented in this dissertation; a) F-LSM demonstration system; b) F-LSM demonstration system with new custom-designed scan lens and; c) miniaturized design including custom-designed scan and tube lens in combination with an alternative 40x CFI S Plan Fluor ELWD objective (Nikon, Japan)

The resulting layout, including the new tube lens and scan lens design in combination with new objective that was simulated as an ideal lens is shown compared to the original optical set up of the demonstration system in figure 5.5. As illustrated in the figure, the optical path length and therefore the size of the revised optical system as measured from the MEMS mirror to the objective, is reduced significantly from a distance of 32 cm to 18 cm. Moreover, the field of view for this new design is 750 µm × 750 µm and thus is 2.34 times as large as the field of view of the original demonstration system. This

95

5 Transfer into a Handheld Microscope Design

improvement in the field of view is obvious when comparing the Strehl Ratios for the different scan angles of the mirror depicted in figure 5.6 a).

a) Strehl Ratio

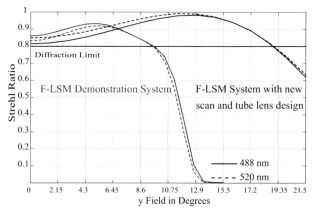

b) Optical Path Difference

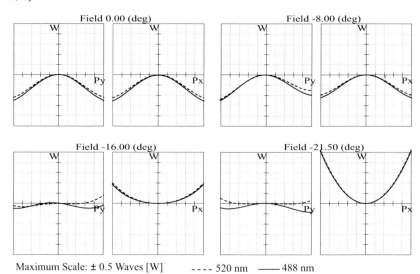

Maximum Scale: ± 0.5 Waves [W]   ---- 520 nm   —— 488 nm

Fig. 5.6: a) Strehl ratio and b) wavefront aberration at the exit pupil of the objective for four different field angles

5.1 Optimized Imaging Optics

It should be noted here that an optical field angle of 15° corresponds to a mirror scan angle of 7.5° as explained in the chapter three. Furthermore, is the maximum effective optical angle, which corresponds to the illumination of the corner of the field of view, given by $\sqrt{2} \cdot 15°$. In consequence it is important to analyze the simulation results up to the field angle of 21.5°. When comparing both point spread functions shown in figure 5.7 of the demonstration system and the new design for on axis situation the optical resolution from both systems is in the range of 1 μm. The OPD is shown for the different field angles in the illustrations below. Here, field curvature is the dominant aberration for large field angles, with some astigmatism present as residual aberration for angles greater than 16°. The corresponding field curvature is less than 4 μm for field angles of 21.5° as depicted in figure 5.7. This is due to the relatively short focal lengths of both the scan and tube lenses. Nevertheless, this curvature does not significantly affect image quality due to the self-correcting effect that only information from the focal plane passes through the pinhole as discussed more in detail in chapter three.

To summarize, the near diffraction-limited optical performance up to field angles of 20° allows adequate cell imaging over a large field of 0.75 mm × 0.75 mm. In comparison to the demonstration system and endoscopes available on the market, the field of view is more than double, making the MEMS-based microscope very promising.

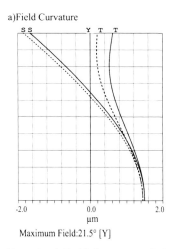
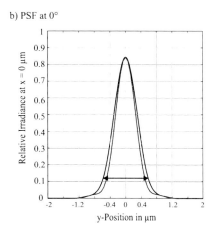

Fig. 5.7: a) Field Curvature of the new design b) point spread function of the demonstration system and the new design for on-axis situation

5 Transfer into a Handheld Microscope Design

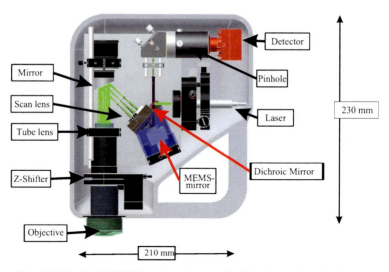

Fig. 5.8: Handheld F-LSM system design with the schematic optical train

5.2 Mounting and Assembly Tolerances

To realize a handheld microscope system for medical applications, a compact solution for integrating the optical systems needs to be found. In addition to achieving a compact layout, mounting and fabrication tolerances are also of great importance for manufacturing and need to be considered in system layout. The resulting theoretical design is shown in figure 5.8. Here, several alignment options are included to ensure the manufacturability as well as optical performance close to the simulated performance above. In order to specify the tolerances, the whole microscope set-up including lens mounts and folding mirrors are investigated and taken into account in the tolerance analysis.

The scan lens, MEMS mirror and dichroic beam splitter are combined within one assembly for space saving purposes, as depicted in figure 5.8. Here, the scan lens barrel is placed into a specially designed mount with a sliding fit to allow axial adjustment of the scan lens to ensure that the mirror is located in the front focal plane of the lens. In addition to holding the lens barrel, this mount also acts as the mount for the dichroic mirror. Using an adapter, screws, and alignment guide rails, this mount is fixed on the MEMS mirror mount, that itself consists of an x-y mount in combination with an adapter onto which the MEMS mirror module is placed. Here the x-y mount is used to account for lateral mounting tolerances of the MEMS mirror on the mirror module as described in chapter four.

5.2 Mounting and Assembly Tolerances

Tip-tilt adjustment is included in the collimator mount to ensure the laser beam is propagating parallel to the optical axis through the scan lens. In addition to this tip-tilt, adjustments for additional x-y adjustment need to be included due to the fact that the tip-tilt mount can compensate for angular misalignments, but does not allow any correction of lateral decentration of the laser beam from the optical axis. For a not any longer tolerable decentration d of 0.2 mm in y-direction the maximum allowable angular misalignment ϕ_{y1}, ϕ_{y2} of the dichroic- and MEMS mirror surface normal relative to the angle bisector of the optical path has been estimated. Without decentration the surface normal of both, the dichroic- and MEMS mirror surface should be located in the xz-Plane ($\phi_{y1} = \phi_{y2} = 0$). The maximum error of Δy is calculated as follows [101]:

$$d = \tan(\phi_{y1}) \cdot l_1 + \tan(\phi_{y1} + \phi_{y2}) \cdot l_2 = 0 \tag{5.4}$$

$$\Delta d = \left| \frac{\delta d}{\delta \phi_{y1}} \right| \cdot \Delta\phi_{y1} + \left| \frac{\delta d}{\delta \phi_{y2}} \right| \cdot \Delta\phi_{y2} + \left| \frac{\delta d}{\delta l_1} \right| \cdot \Delta l_1 + \left| \frac{\delta d}{\delta l_2} \right| \cdot \Delta l_2 \tag{5.5}$$

The distances l_1, l_2 refer to those denoted in figure 5.9 and $\Delta\phi_{y1}$ and $\Delta\phi_{y2}$ are the maximum errors for ϕ_{y1} and ϕ_{y2}. With $\Delta\phi_{y1}$ and $\Delta\phi_{y2}$ assumed to be of equal amount $\Delta\phi$, this leads to:

$$\Delta y = (l_1 + l_2 + l_2) \cdot \Delta\phi \tag{5.6}$$

For $l_1 = 22$ mm, $l_2 = 30$ mm and the above stated, not any longer tolerable, decentration Δd of 0.2 mm, this would require the maximum error $\Delta\phi$ to be smaller than 0.0015.

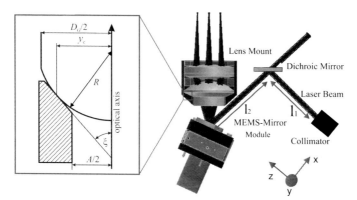

Fig. 5.9: Lens mount with conical contact mount for the first doublet inside the drop-in mount

5 Transfer into a Handheld Microscope Design

As a consequence, an additional lateral adjustment option for the collimator mount needs to be included. For this reason a 4-axis mount providing tip, tilt, x- and y- adjustments was chosen.

For the scan lens mount, mounting and manufacturing tolerances have to be addressed separately. Manufacturing tolerances such as surface centering errors were discussed with an optical supplier and are listed in Apendix B. For lens mounting a "drop-in" design concept was used with both doublets being separated by a ring spacer. This spacer is a critical part of the mount since it needs to be thin, of correct diameter and precision fit in the bore to avoid the lens to tilt and to allow self-centering to take place. Self-centering refers to the situation, that an applied axial preload will tend to move the lens toward the centered condition when the following equation involving the heights of contact y (measured from the optical axis) and radii R of each surface is satisfied [102]:

$$(2y/R_4) + (2y/R_3) >= 4\eta = 0.56 \tag{5.7}$$

where $\eta = 0.14$ is a typical coefficient of friction applicable to glass sliding on polished steel. Applying equation 5.7 for the scan lens with $R_3 = 16.55$ mm and $R_4 = 35.69$ mm leads to:

$$y \geq \frac{2\eta R_3 R_4}{R_3 + R_4} = 3.17 \text{ mm} \tag{5.8}$$

Thus, the self-centering condition is fulfilled for the used spacer with a clear aperture of 16.6 mm. To maintain a large aperture and to reduce the contact stress on the first element inside the drop-in mount, a conical contact mount has been chosen. This tangential interface with the lens is regarded as nearly the ideal interface for convex surfaces and can be fabricated fairly easily [103]. With an overall lens diameter of $D_{Lens} = 22$ mm and a clear aperture of $A = 14.4$ mm of the mount, the midpoint between this clear aperture and the rim of the lens is given by [103]:

$$y_c = (D_G + A)/4 = 9.1 \text{ mm} \tag{5.9}$$

To achieve the lens being held at y_c, the cone half-angle ξ needs to be:

$$\xi = 90° - \arcsin(\frac{y_c}{R_2}) = 65° \tag{5.10}$$

Both doublets and the spacer fit into the inner diameter of the drop-in mount with 0.075 mm clearance of the diameter. A threaded ring retainer holds the whole assembly securely.

The illumination path includes a fold mirror, the tube lens, the objective, and the z-shifter. Here a tubus system was applied for stray light reduction and a prism holder

5.2 Mounting and Assembly Tolerances

providing tip-tilt adjusted was used for the fold mirror. The alignment tolerances of these components are the same as discussed in chapter three for the demonstration system.

A special design is used for the interface of the microscope objective with the sample that is easily interchanged and cleaned for hygienic reasons. Moreover, the interface as depicted in figure 5.10 ensures that there is no direct contact between the sample and the objective lens. This is especially important considering the working distance (3 mm) and the intended application in a clinical setting. The end cap incorporates a 0.5 mm-thick sapphire window for this reason. This sapphire window can cause additional spherical aberration which should be tested upfront due to missing objective data. The cap itself is axially screwed on a 40 mm-outer-diameter hollow, cylindrical attachment at the housing which has a M38×0.5 mm internal thread. A retainer ring with a rubber joint outside the cylindrical attachment defines the depth to which the end cap can be screwed in without hitting the objective's front lens and impedes ingress of impurities.

The entire detection unit resembles the one in the demonstration system to a great extent. It is based on the tube system from Qioptiq with a spatial filter module to provide precise xy-adjustment for the pinhole, as well as adjustment along the optical axis. The focusing lens and an additional filter is positioned in in close proximity to the front of a right-angled cage mount for mirrors mounted at a 45° angle within a tube system. This kinematic mount secures the optic inside and allows ±4° of adjustment using threaded adjusters. The coupling with the detector is provided by a C-mount thread.

All these different parts of the microscope are mounted on a chassis plate and placed in a specially designed housing depicted in figure 5.10.

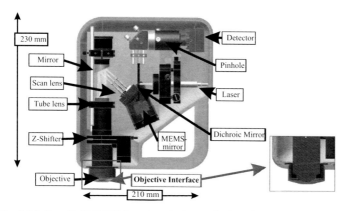

Fig. 5.10: Handheld F-LSM system design with the schematic optical train

5 Transfer into a Handheld Microscope Design

To carry out the final tolerance analysis of the imaging system, the tolerances above discussed were applied using an iterative process of design and tolerance analysis. The system having a 'realistic' chance being properly manufactured was defined based on experience at an 80%-Monte Carlo level with Strehl ratio higher than 0.7 for field angles up to 20°. The final set of manufacturing and mounting tolerances including the compensators, which are the image plane distance and the distance between the tube and scan lens, are listed in table 5.2.

Tab. 5.2: Tolerance Assumptions for the Monte Carlo Analysis

Tolerances on	Value ($\pm$)	Unit
Radius of Curvature		
All Lenses	0.4	%
Lens Element Thickness		
All Lens Elements	30	μm
Element Distance		
Eye Relief	200	μm
Scan Lens Elements 1 & 2	50	μm
Tube Lens - Objective	800	μm
Element Decenter		
Scan Lens Element 1,2	50	μm
Tube Lens	75	μm
Element Tilt		
Scan Lens Element 1,2	0.05	°
Tube Lens	0.1	°
Surface Decenter		
All Lens Surfaces	30	μm
Surface Tilt		
All Lens Surfaces	0.033	°
Index		
All Glasses	0.0005	
Abbe Number		
All Glasses	0.8	%
Distance Compensator 1	Scan Lens - Tube Lens	
Distance Compensator 2	Image Distance	

5.2 Mounting and Assembly Tolerances

Figure 5.11 shows the results on the Strehl ratio of one hundred Monte Carlo trials superposed on each other. The data prove the system meets the design criterion in more than 80% of the Monte Carlo trials. With the sensitivity analyisis the maximum change in the systems performance is in the first step evaluated for each tolerance operand individually an then in a second step adjusted by compensation elements to achieve a final measure for the sensitivity of the different tolerance operators. In the prescent system the distance tolerance between the MEMS mirror and the scan lens with ±200 µm and the decentration tolerance of the tube lens (±75 µm) were the tolerance opterands of highest sensitivity. In summary the final tolerance analysis has confirmed that it is feasible to manufacture the new optical design.

a) Strehl Ratiofor 100 Monte Carlo Trials during Tolerance Analysis

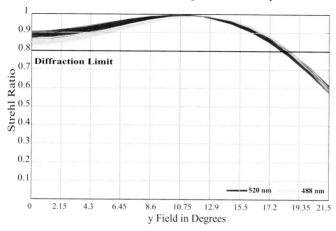

Fig. 5.11: Strehl ratio of the final imaging optics design as a function of the field angle for one hundred Monte Carlo trials superposed on each other

5.3 Summary and Comparison of Both F-LSM Systems

To summarize, the transformation of the demonstration system into the design of a handheld version by exploring and designing a new optical imaging system demonstrates that the MEMS-based approach allows the realization of a handheld F-LSM system with extended field of view. In the theoretical handheld system depicted in figure 5.12, the optical imaging system consists of a new scan and tube lens design consisting four-element and three-element lenses respectively, in combination with a 40× CFI S Plan Fluor objective (Nikon, Japan). These newly developed lenses produce a large improvement in the maximum field of view, and the shorter optical path length of the imaging system reduces the size of the overall layout. The optical path length measured from the MEMS mirror to the objective is reduced by 140 mm from the initial 320 mm path length, the field of view is enlarged by more than a factor of two, while the lateral and axial resolutions remain unchanged to allow adequate cell imaging. The enlargement of the field of view in the handheld system was possible due to the fact the demonstration system could not use the maximum possible MEMS mirror scan angles because of limited optical performance of the imaging system. By integrating custom-designed lenses, the simulation results indicate diffraction-limited performance for mirror scan angles up to $7.5°$, resulting in a field of view measuring 750 µm × 750 µm. This field of view is more than two times larger than the original field of view of 490 µm×490 µm. Considering that the largest field of view for endoscopes available on the market is even less that of the demonstration system, this is a very promising result for the future application of MEMS mirrors in such systems.

5.3 Summary and Comparison of Both F-LSM Systems

F-LSM Demonstration System

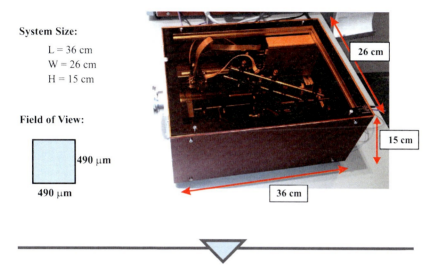

System Size:
L = 36 cm
W = 26 cm
H = 15 cm

Field of View: 490 µm × 490 µm

New F-LSM Design Based on Customized Optics

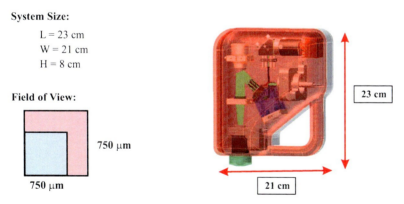

System Size:
L = 23 cm
W = 21 cm
H = 8 cm

Field of View: 750 µm × 750 µm

Fig. 5.12: Photograph of the initial F-LSM system introduced in chapters three and four in comparison to the CAD graphic of the handheld F-LSM design based on optimized designed imaging optics

6 Discussion and Outlook

Fluorescence confocal laser scanning microscopes are well established and widely used in fundamental biological and medical research. In contrast to optical wide-field microscopes, confocal fluorescence microscopes can perform depth-resolved fluorescence imaging of 'thick' specimens with increased imaging contrast by effectively suppressing out-of-focus light. Due to this ability of surveying cross-sections of skin in a non-invasive manner, recent research activities in the field of dermatologic diagnostics have proven F-LSM to be a highly promising tool in clinical dermatology. It becomes possible with this imaging technique to qualitatively evaluate dynamic skin processes and to perform penetration and distribution studies of topically applied substances in vivo. The only preparation needed for these measurements is the application of a fluorescent contrast agent in advance to label the skin structure.

However, despite these benefits, which can lead to a reduction of performed biopsies, fluorescence confocal laser scanning microscopes are uncommon thus far in medical practice. Factors which have impeded its establishment are the small field of view and large size of the F-LSM systems dealt with in medical and biological research. For dermatology, it would be desirable to a have a compact system with a large field of view and a resolution appropriate for cell imaging.

Several approaches have been pursued within the last ten years to miniaturize F-LSMs. Among them, fiber scanning endoscopes have been prominent examples designed for in-vivo imaging inside the human body. They necessarily possess a small field of view. However, a large field of view is of interest in dermatologic diagnostics, whereby a handheld system layout is sufficient.

Hence, the goal of this research project was to investigate a new approach that integrates a MEMS mirror for laser scanning. With a first compact microscope system for functional verification, already a field of view larger than that of currently available endoscopes was demonstrated. To illustrate the optical performance, several depth-resolved images of histological and biological samples were presented. It was also shown that an optimized optics design will allow to transform the microscope into a handheld version with a more than twofold enlarged field of view. In summary, this research project showed that the MEMS-based approach is a highly promising solution thanks to its distinctive ability to provide a large field of view while allowing a compact layout.

The major challenges associated with the research and realization of the microscope system were (1) using a dual-axis MEMS mirror (provided by the Fraunhofer IPMS) to

6 Discussion and Outlook

create a compact scanning mechanism, (2) creating an affordable and highly corrected optical imaging system that delivers diffraction-limited performance for the fluorescence excitation and emission wavelengths over a large field of view, (3) suitable detection optics and filters to ensure depth-resolved confocal imaging, and (4) realization of a space-saving and compact microscope system.

The final realization process involved first the definition of target specifications and study of the resulting implications and requirements for the optical layout and detection unit resulting from the MEMS mirror integration. Based on these results, the microscope optics was designed and optimized with the help of commercially available optical design software. To correct chromatic aberrations while keeping the cost for the first F-LSM system low, off-the-shelf achromatic doublets and a chromatic-corrected Hastings Triplet were used in the design. The final tolerance analysis involved defining necessary mounting for adjustment options to ensure adequate optical performance in spite of manufacturing and mounting tolerances. A special target holder to provide lateral and axial adjustment options for the sample was designed for the final assembly of the first F-LSM set-up.

The MEMS-based confocal fluorescence microscope system in its final configuration measures ($26\,cm \times 15\,cm \times 34\,cm$ (w x h x d)) and consists mainly of off-the-shelf components with total fabrication and assembly costs of less than 40,000 EUR. With a measured lateral and axial resolution of around 2 µm, the $0.49\,mm \times 0.49\,mm$ field of view limited by the optical performance of the catalog lenses use, but is nevertheless larger than those of commercially available endoscopes exhibiting comparable resolution. Various measurements on biological samples illustrated the sectioning capabilities and adequate resolution for cell imaging in dermatology with this resolution and field of view.

For quantitative evaluation, microscope performance test measurements on selected samples and wavefront error measurements with a Shack-Hartmann were performed. It could be shown through this combined evaluation that the lateral resolution was not limited by the optics, but instead by the calculation of the laser beam position used for assigning a measured intensity value to a certain pixel. The accuracy of this calculation depends on the degree of correspondence between the stored oscillation function of the mirror and the actual performance with which the laser beam position in the image field is calculated. Furthermore, the Shack-Hartmann measurement identified a slight MEMS mirror curvature that leads to typical astigmatic deviations of the wavefront in the exit pupil of the microscope.

The objective has been to transform the initial F-LSM system into a handheld F-LSM design delivering an enlarged field of view and comparable optical resolution. The objective was achieved by exploring, investigating, and designing a optimized optical imaging system. Here, a two-step approach was taken. First, the scan lens was exchanged for a custom-designed, four-element lens system. Then a new custom-designed tube lens in combination with an alternative objective were included. Special attention was paid to the scan lens design due to its remote aperture position, which in combination with large scan angles complicates the correction of odd wavefront aberrations. The complete microscope design has been simulated and analyzed by for residual aberrations and mounting tolerances using optical design software. Boundary conditions, variables and glass substitution templates were used during this process. The resulting theoretical design of a handheld F-LSM system using the new optics design exhibits more than a two-fold enlargement of the field of view measuring 0.75 mm × 0.75 mm, It also exhibits more than 40% reduction in optical path length. Considering that the largest field of view for endoscopes available on the market is even less that of the initial F-LSM system, this a promising result for the future application of MEMS mirrors in such systems.

For applications where a resolution higher than the 2 µm is necessary, mirror position assessment needs to be improved. Several alternative approaches are currently being investigated at Fraunhofer IPMS to overcome this restriction that results from the deviation of the approximated sinusoidal mirror oscillation and its real form. These approaches include improved approximation functions, look-up tables, and different direct measurement methods. Initial tests are being carried out. As a result, improved assessment of the mirror position is likely to be implemented in the next laser scanning devices currently under development. In addition the integration of linear mirrors is conceivable, which would reduce the image acquisition time significantly. The development of a one-dimensional linear-drive MEMS mirrors designed for close-coupled configurations is nearing completion. These mirrors are likely to deliver shorter image acquisition times with improved image resolution. Image resolution would then be determined by optical performance rather than the determination of mirror position.

The current resolution is adequate for cell imaging in dermatology. It might be useful in this application to include options in the instrument such as pre-display of the region which will be imaged using some sort of crosshairs or image boundary indicators. This could be achieved by coupling light from a second light source into the optical beam path, either in front of the mirror or behind it between the tube lens and the objective. It will then be possible to project crosshairss or any other figures onto the skin through adequate light modulation. This would also allow localized real-time treatment of the skin using certain wavelengths if required.

Very promising results have recently been achieved in the design of substances that penetrate the outermost skin layer of the stratum corneum. Combining fluorescence dyes with such substances would eliminate the need to inject the fluorescence dye. True non-invasive in-vivo assessment of the skin will then be possible.

In view of current research activities in addition to the existing medical applications, the application of F-LSM systems in medical diagnostics shows large potential. MEMS-based systems in particular have a good chance to become established in practice thanks to their distinctive ability to provide a large field of view while being compact, as demonstrated in this research project.

Appendix A: Flow Chart

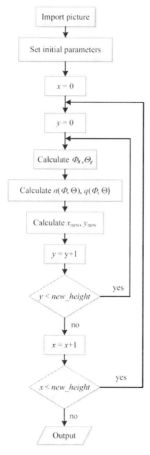

Fig. 1: Flow Chart of the Program in Section 3.5

Appendix B: Scan Lens - Technical Drawings

The technical drawing of the scan and tube lens design was generated in accordance with ISO standard 10110. This ISO standard is the primary reference for technical drawings and specifies the presentation of design and functional requirements for optical elements and systems in technical drawings [[104],[105]]. It consists of 13 sections that each describe a different aspect of the complete drawing [106]. Table 1 give a partial listing of the ISO 10110 specifications used for the optical drawings. The technical drawing of the scan lens design was generated in accordance with these specifications. The optical requirements and tolerances included in the technical drawing of the scan lens shown in figure 2 are typical quality indications, as discussed with the optics supplier.

Tab. 1: ISO 10110 specifications used for the technical drawing Fig. 2

Indication	Parameter	Comment
0/ 10	Stress birefringence	Maximum OPD is 10 nm/cm
1/ 3 x 0.1	Bubbles and inclusion	Allow up to 3 inclusion, each no longer than 100 μm over the optical clear aperture
2/ 2:3	Inhomogeneity and Striae	Homegoneity class 2 corresponds to $\Delta n = \pm e^{-6}$, Striae class 3 is $< 2\%$
3/ 0.25 (.2/.125)	Surface form error	Fringe and optical path difference given for $\lambda = 546.07$ nm: 0.25 fringe of sag(power)error, a maximum p-v irregularity of 0.2 fringe and 0.125 fringe of rotational symmetric irregularity error
4/ 1.5' (0.033)	Centering error	Element wedge is 1.5 arc minute
5/ 5x0.05; C3x0.16;E0.5	Surface form error	allow up to 5 digs, each no larger than 50 μm, allow 3 coating imperfection no larger than 160 μm and 1 edge chip no larger than 0.5 mm

Appendix B: Scan Lens - Technical Drawings

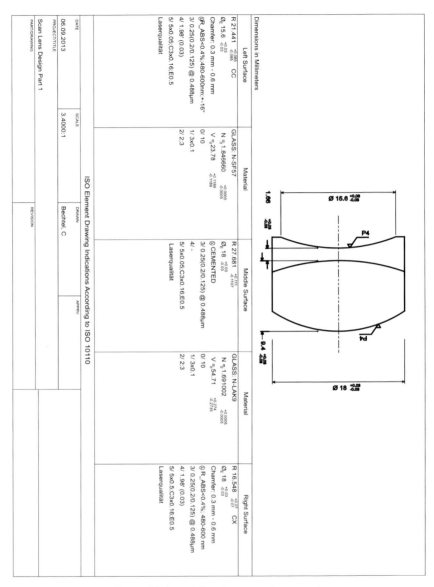

Fig. 2: Technical drawing - scan lens part 1

114

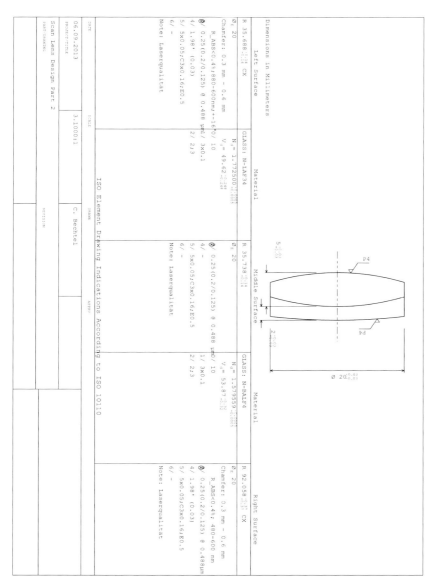

Fig. 3: Technical drawing - scan lens part 2

Abbreviations

5-ALA	5-Aminolevulinic Acid
AFL-HSA	5-aminofluorescein-human Serum Slbumin
A/D converter	Analog-to-Digital Converter
BfArM	Bundesinstitut für Arzneimittel und Medizinprodukte
	Federal Institute for Drugs and Medical Devices
CCD	Charge-coupled Device
FEA	Finite Element Analysis
F-LSM	Fluorescence Confocal Laser Scanning Microscope
FPGA	Field Programmable Gate Array
ICG	Indocyanine Green
IPMS	Fraunhofer Institute for Photonic Microsystems
LED	light-emitting Diode
MEMS	Micro Electro Mechanical System
MTF	Modulation Transfer Function
NA	Numerical Aperture
N_A	Avogadro Constant
NEP	Noise Equivalent Power
OCT	Optical Coherence Tomography
OPD	Optical Path Difference
OTF	Optical Transfer Function
PDE	Photon Detection Efficiency
PpIX	Protoporphyrin IX
PSF	Point Spread Function
P-V value	Peak-to-Valley Value
PZT	Lead Zirconium titanate
R-LSM	Reflectance Confocal Laser Scanning Microscopy
RMS	Root Mean Square
SNR	Signal-to-Noise Ratio
SOI	Silicon on Insulator
TEWL	Transepidermal Water Loss
USAF	United States Air Force

List of Variables

$\Delta\phi_x$	smallest detectable angles in scan direction x
$\Delta\phi_y$	smallest detectable angles in scan direction y
$\Delta\phi$	diffracted angle
δ_{mirror}	mirror height deformation
ϵ	extinction coefficient
ϕ	mirror scan angle
ρ	normalized pupil height
σ	absorption cross-section (area)
τ	fluorescence lifetime
Θ	total optical scan angle
ΘD	ΘD-product
ω_0	spot radius ($1/e^2$)
Ω	solid angle
A	focal spot area
a_1	defocus coefficient
a_2	tilt coefficient
b_1	coefficient for spherical aberration
b_2	coefficient for coma
b_3	coefficient for astigmatism
b_4	coefficient for field curvature
b_5	coefficient for distortion
c_{dye}	fluorophore dye concentration
D	diameter
D_{Airy}	diameter of the Airy disk
D_{eff}	effective mirror diameter
D_{mirror}	mirror diameter
d_{im}	length of one side of the square field of view
E_{Photon}	energy of a photon
f	focal length
f_{fast}	scanning frequency of the fast axis of the mirror
f_{slow}	scanning frequency of the slow axis of the mirror
$f/\#$	f-number
f_{scan}	focal length of the scan lens

List of Variables

f_{tube}	focal length of the tube lens
F_{max}	maximum possible fluorescence photon emission rate
G	gain
$G_{preamplifier}$	preamplifier gain
$G_{internal}$	internal gain of the photodetector
H	normalized image height
I_N	resolution invariant
I_{sat}	required incoming photon flux
k	aperture shape factor
M	magnification
N_A	Avogadro's Number
N_{Theta_x}	total number of spots in direction x
N_{Theta_y}	total number of spots in direction y
N_{Photon}	number of photons received per pixel
N_{dye}	number of dye molecules within the focal volume
n	number density of the fluorescein molecules
n_1, n_2	refractive indices
NA	numerical aperture
NA_{pin}	numerical aperture of the pin hole
NA_{obj}	numerical aperture of the objective
P	optical power
P_{laser}	laser power
PDE	photodetector efficiency
QY	quantum yield
r_0	entrance aperture radius
$r_{aperture}$	aperture radius
r_{beam}	Gaussian beam radius
r_{eff}	effective mirror radius
S	photodetector responsivity
t	illumination time
T	transmission
V	focal volume
$W(\rho, \Theta)$	wave aberration function
W_{rms}	root mean square error on the wave-front
z_{int}	shift of the intermediate image plane

List of Figures

2.1 Optical path of a) a confocal microscope with stage scanning per Minsky, 1957, b) confocal laser-scanning microscope using two scanning mirrors . 6
2.2 Configurations of confocal microscope scanning system using galvanometric scanning mirrors [24] 7
2.3 Gimbal-mounted dual-axis MEMS mirror with comb-drive actuators [28] 8
2.4 Fiber scanning techniques based on a) PZT scanning per [37] and b) tuning fork scanning per [38] 10
2.5 a) Application of a fluorescent contrast agent to label the cells for displaying skin structure and b) F-LSM images of different skin layers ([44],[42]) 11
2.6 F-LSM images of the typical epidermal wound healing process taken by Lademann et al. 12
2.7 One form of a Jablonski diagram showing the processes involved in the decay of a molecule with a ground state and excited state 14
2.8 Absorption and emission spectra from fluorescein [64] and its corresponding chemical structure (right)[65] 16
2.9 Schematic of an optical imaging system with an exaggerated illustration of an aberrated wavefront for an off-axis point object 20
2.10 Illustration of a) spherical aberration [70] and b) coma [70] 22
2.11 Illustration of a) astigmatism and b) field curvature [70] 22
2.12 Illustration of a) different kind of image distortions [70] and b) lateral chromatic aberration [70] 23
2.13 Relation between the Pupil function, PSF, Strehl ratio, MTF and OTF of an imaging system 25
3.1 a) Schematic optical train of the fluorescence confocal laser scanning microscope based on a dual-axis MEMS mirror, b) photograph of the dual-axis MEMS mirror module with integrated electronic interface (Fraunhofer IPMS) 29
3.2 Total number of resolvable spots of a scanner system 30
3.3 Illustration of the Lagrange invariant (image from [75]) 33
3.4 Confocal detection principle from [67] 35

LIST OF FIGURES

3.5 Schematic of the optical train of the F-LSM based on a dual-axis MEMS mirror where the imaging optic simulated in the design process are highlighted 39

3.6 Optical layout with rays traced for three different deflections of the dual-axis MEMS mirror. The objective lens is modeled as an ideal lens 40

3.7 Predicted performance of the optical design in figure 3.6 a) Huygens PSF for four different field angles b) Strehl Ratio as a function of the field angle 41

3.8 a) Wave aberration at the exit pupil of the objective for four different field angles b) Field curvature in tangential and sagittal directions 42

3.9 Strehl ratio as a function of the field angle for a system arrangement generated by the Monte Carlo analysis with an optical performance worse than 80% of all generated system arrangements 46

3.10 Dynamic MEMS mirror deformation with the expression of the deformation by the first ten Zernike polynomial coefficients for a scan angle of 7.5°(FEA simulation performed by a colleague from Fraunhofer IPMS [82]) 47

4.1 a) Principle set-up: laptop, remotely connected power and control units and the microscope system with the schematic optical train b) Photograph of the MEMS-based F-LSM demonstration system 56

4.2 Computer-aided design of a) the five-axis mount fastened on the back side of the microscope b) x-y adjustment set-up for the dual-axis MEMS mirror and c) the mirror mount 58

4.3 a) Photograph of the detection unit and b) the transmission spectra of the dichroic and band-pass filters, respectively 59

4.4 a) Photograph of the target slide holder during illumination and b) the corresponding computer-aided design of the holder with its matching protection cover for trade fairs. 60

4.5 Principal set-up: laptop, remotely connected power and control unit, and the microscope system with the integrated electronics module 61

4.6 a) Schematic optical train of the calibration set-up b) illustration of the effect when the real and stored mirror position functions differ and c) schematic illustration of this phase differences on the appearance of a pinhole image 62

4.7 Photograph of the F-LSM demonstration system 63

4.8 With a Varian Cary spectrometer measured fluorescence spectrum of the Fluor-Ref fluorescent slide upon excitation with 488 nm (right) and photograph of the USAF resolution target mounted on top of the fluorescent slide(left) 64

LIST OF FIGURES

4.9   a) Test set-up for the microscope objective to ensure the feasibility of bringing the chrome structure in the focal plane of the objective, the images of a 1 mm test structure are displayed in b) and c) for the two possible orientations 65

4.10 Images obtained with the F-LSM demonstration set-up a) USAF resolution target b) scaled reticle with 1 mm scale and c) microring resonator stained with 1 µm diameter microspheres (F-13081; Molecular Probes) . 66

4.11 a)Illustration of a fluorescent microsphere with the corresponding fluorescence spectrum (right)[65] 67

4.12 Serial optical sectioning from top to bottom along the z-axis of ring-stained microspheres reveal the optical sectioning capabilities of the F-LSM demonstrator. The diameter of the fluorescent ring (or disc) seen is dependent on the depth of the optical focal plane. 68

4.13 Images of the upper fluorescent bead in two different sectioning depths with corresponding intensity distributions (right) 68

4.14 a)Δh for eleven measurements (left) and a stack of intensity profiles of the upper bead in figure 4.12 along the diameter of the circle axial (right); b) measurement uncertainty calculation result with Mathematica for h with $\Delta s = 2$ µm and $\Delta r = \pm 0.07$ µm[65] 69

4.15 Wide-field microscopy (Keyence, Digital Microscope VHX 2000) image of a v-groove filled with microspheres 70

4.16 Images of a V-groove etched in silicon and filled with 0.4 µm fluorescent microspheres a) and b) with the software ImageJ created stack of images that were taken with the F-LSM system in different focal depths [90] c) - e) images of the v-groove taken at different focal depths 71

4.17 Basic layout of a Shack-Hartmann wavefront sensor 72

4.18 Photograph of the wavefront measurement set-up with the Shack Hartmann sensor 73

4.19 a) Shack-Hartmann wavefront measurement result in terms of the wavelength λ, b) reconstruction of the wavefront measurements from a) above using Zernike polynomials on the left and the corresponding Zernike coefficients on the right c) P-V wavefront measurement variations from 174 reference measurements, and d) the P-V wavefront deviations from the fourteen final measurements 75

4.20 a) Interferometric measurement and reconstruction of a dual-axis MEMS mirror and measurement results of the x and y profiles in b) and c) revealing mirror curvature of 1.5 m 78

4.21 Comparison of the Shack-Hartmann measurement result a) with the simulation results b). The corresponding OPD graphs are shown in c). . . . 79

4.22 Fluorescence spectrum and photograph of a stained histological skin sample 80

LIST OF FIGURES

4.23 Images from sections of a stained histologic skin sample taken with the F-LSM demonstration set-up; (a) section of the epidermis and the dermis with the different layers labeled; (b) longitudinally cut profile of a hair follicle with sebaceous glands; (c)-(e) different appendages of the skin (f) cut profile of sweat glands embedded in the connective tissue of the dermis 81

4.24 Series of images taken with the F-LSM demonstration system of an acridin orange stained convalaria stem (depths difference between the different images: 3 μm) 82

4.25 Images of "thick" cross-sectional samples from different plants; a)F-LSM image of a tensile flower (b)wide-microscopy image of tensile c)-d) F-LSM images of different sections from a convalaria sample; e)-f) cross-sectional F-LSM images of lavender 83

4.26 Photograph of the F-LSM demonstration system on a trade fair 85

5.1 Schematic of optical train of the F-LSM system based on a dual-axis MEMS mirror where optical elements of the imaging system are highlighted 88

5.2 Basic forms of three- and four-element optical eyepieces (data from [98]) 90

5.3 a) Optical layout including the newly designed scan lens; b) scan lens specifications; c) Strehl ratio of the original F-LSM demonstration system in comparison to the new design 92

5.4 Wavefront aberrations at the exit pupil of the objective for four different field angles 93

5.5 Illustration of the different optical layouts presented in this dissertation; a) F-LSM demonstration system; b) F-LSM demonstration system with new custom-designed scan lens and; c) miniaturized design including custom-designed scan and tube lens in combination with an alternative 40x CFI S Plan Fluor ELWD objective (Nikon, Japan) 95

5.6 a) Strehl ratio and b)wavefront aberration at the exit pupil of the objective for four different field angles 96

5.7 a) Field Curvature of the new design b)point spread function of the demonstration system and the new design for on-axis situation 97

5.8 Handheld F-LSM system design with the schematic optical train 98

5.9 Lens mount with conical contact mount for the first doublet inside the drop-in mount 99

5.10 Handheld F-LSM system design with the schematic optical train 101

5.11 Strehl ratio of the final imaging optics design as a function of the field angle for one hundred Monte Carlo trials superposed on each other . . . 103

LIST OF FIGURES

5.12 Photograph of the initial F-LSM system introduced in chapters three and four in comparison to the CAD graphic of the handheld F-LSM design based on optimized designed imaging optics 105

1   Flow Chart of the Program in Section 3.5 111
2   Technical drawing - scan lens part 1 114
3   Technical drawing - scan lens part 2 115

List of Tables

2.1	Target Specifications	13
2.2	Fluorophores for human use in medicine (sources: [55],[56],[57],[58],[59])	15
2.3	Characteristic Properties of Fluorophores (sources: [66],[67],[26])	18
3.1	Intensity losses at the different elements in the detection path	37
3.2	F-LSM System Specifications	40
3.3	Components	40
3.4	Optical System Specifications for the Initial F-LSM Layout	42
3.5	Tolerance Assumptions for the Monte Carlo Analysis	45
4.1	Major components used in the F-LSM Demonstration System Set-up	57
4.2	Axial resolution measurements	70
4.3	Seidel aberrations corresponding to the first Zernike polynomials [[83], [80]]	77
4.4	MEMS mirror curvature: interferometric measurement results	77
4.5	F-LSM Demonstration System Specifications	85
5.1	Optical System Specifications	94
5.2	Tolerance Assumptions for the Monte Carlo Analysis	102
1	ISO 10110 specifications used for the technical drawing Fig. 2	113

Bibliography

[1] M. Minsky. Memoir on inventing the confocal scanning microscope. *Scanning*, 10(4):128–138, 1988.

[2] R. R. Anderson and J. A. Parrish. The optics of human skin. *Journal of Investigative Dermatology*, 77(1):13–19, 1981.

[3] F. Helmchen, M. S. Fee, D. W. Tank, and W. Denk. A miniature head-mounted two-photon microscope: high-resolution brain imaging in freely moving animals. *Neuron*, 31(6):903–912, 2001.

[4] H. Bao, J. Allen, R. Pattie, R. Vance, and M. Gu. Fast handheld two-photon fluorescence microendoscope with a 475 μm× 475 μm field of view for in vivo imaging. *Optics letters*, 33(12):1333–1335, 2008.

[5] C. J. Engelbrecht, R. S. Johnston, E. J. Seibel, and F. Helmchen. Ultra-compact fiber-optic two-photon microscope for functional fluorescence imaging in vivo. *Optics express*, 16(8):5556–5564, 2008.

[6] R. Le Harzic, M. Weinigel, I. Riemann, B. Messerschmidt, et al. Nonlinear optical endoscope based on a compact two axes piezo scanner and a miniature objective lens. *Optics express*, 16(25):20588–20596, 2008.

[7] J. Sawinski and W. Denk. Miniature random-access fiber scanner for in vivo multiphoton imaging. *Journal of Applied Physics*, 102(3):034701–034701, 2007.

[8] C. L. Arrasmith, D. L. Dickensheets, A. Mahadevan-Jansen, et al. Mems-based handheld confocal microscope for in-vivo skin imaging. *Optics express*, 18(4):3805–3819, 2010.

[9] H. Ra, W. Piyawattanametha, M. J. Mandella, P.-L. Hsiung, J. Hardy, T. D. Wang, C. H. Contag, G. S. Kino, and O. Solgaard. Three-dimensional in vivo imaging by a handheld dual-axes confocal microscope. *Optics express*, 16(10):7224, 2008.

[10] L. Fu, A. Jain, H. Xie, C. Cranfield, and M. Gu. Nonlinear optical endoscopy based on a double-clad photonic crystal fiber and a mems mirror. *Optics Express*, 14(3):1027–1032, 2006.

Bibliography

[11] T.-M. Liu, M.-C. Chan, I. Chen, S.-H. Chia, C.-K. Sun, et al. Miniaturized multiphoton microscope with a 24hz frame-rate. *Optics express*, 16(14):10501–10506, 2008.

[12] W. Jung, S. Tang, D. T. McCormic, T. Xie, Y.-C. Ahn, J. Su, I. V. Tomov, T. B. Krasieva, B. J. Tromberg, and Z. Chen. Miniaturized probe based on a microelectromechanical system mirror for multiphoton microscopy. *Optics letters*, 33(12):1324–1326, 2008.

[13] W. Piyawattanametha, E. D. Cocker, L. D. Burns, R. P. J. Barretto, J. C. Jung, H. Ra, O. Solgaard, and M. J. Schnitzer. In vivo brain imaging using a portable 2.9 g two-photon microscope based on a microelectromechanical systems scanning mirror. *Optics letters*, 34(15):2309, 2009.

[14] J. T. C. Liu, M. J. Mandella, N. O. Loewke, H. Haeberle, H. Ra, W. Piyawattanametha, O. Solgaard, G. S. Kino, and C. H. Contag. Micromirror-scanned dual-axis confocal microscope utilizing a gradient-index relay lens for image guidance during brain surgery. *Journal of biomedical optics*, 15(2):026029–026029, 2010.

[15] J. Z. Young and F. Roberts. A flying-spot microscope. *Nature Publishing Group*, 1951.

[16] P. Davidovits and M. D. Egger. Scanning laser microscope for biological investigations. *Applied Optics*, 10(7):1615–1619, 1971.

[17] P. Davidovits. Scanning optical microscope, 1972. US Patent 3,643,015.

[18] K. Carlsson and N. Aslund. Confocal imaging for 3-d digital microscopy. *Applied optics*, 26(16):3232–3238, 1987.

[19] J. G. White. Confocal scanning microscope, December 15 1987. UK Patent 2,184,321.

[20] R. W. Resandt, H. J. B. Marsman, R. Kaplan, J. Davoust, E. H. K. Stelzer, and R. Stricker. Optical fluorescence microscopy in three dimensions: microtomoscopy. *Journal of microscopy*, 138(1):29–34, 1985.

[21] J. G. White. Confocal imaging system, July 16 1991. US Patent 5,032,720.

[22] D. B. Murphy and M. W. Davidson. *Fundamentals of light microscopy and electronic imaging*. Wiley-Blackwell, 2012.

[23] R. H. Webb. Confocal optical microscopy. *Reports on Progress in Physics*, 59(3):427, 1996.

[24] E. H. K. Stelzer. *The intermediate optical system of laser-scanning confocal microscopes*. Springer, 2006.

[25] J. Rietdorf and E. H. K. Stelzer. *Special optical elements*. Springer, 2006.

[26] J. Pawley. *Handbook of Biological Confocal Microscopy*, volume 3 of *Language of science*. Springer, 2006.

[27] K. R. Spring, T. J. Fellers, and M. W. Davidson. Confocal microscope scanning systems. Technical report, Olympus Corporation,, http://www.olympusconfocal.com/theory/confocalscanningsystems.html., 2009.

[28] S.-T. Hsu, T. Klose, C. Drabe, and H. Schenk. Two dimensional microscanners with large horizontal-vertical scanning frequency ratio for high-resolution laser projectors. pages 688703–688703, 2008.

[29] H. Schenk, P. Dürr, D. Kunze, H. Lakner, and H. Kück. A resonantly excited 2d-micro-scanning-mirror with large deflection. *Sensors and Actuators A: Physical*, 89(1):104–111, 2001.

[30] M. Scholles, A. Bräuer, K. Frommhagen, C. Gerwig, H. Lakner, H. Schenk, and M. Schwarzenberg. Ultracompact laser projection systems based on two-dimensional resonant microscanning mirrors. *Journal of Micro/Nanolithography, MEMS, and MOEMS*, 7(2):021001–021001, 2008.

[31] U. Schelinski, J. Knobbe, H. Dallmann, H. Grüger, M. Förster, M. Scholles, M. Schwarzenberg, and R. Rieske. Mems based laser scanning microscope for endoscopic use. In *Proc. of SPIE Vol*, volume 7930, pages 793004–1, 2011.

[32] E. J. Seibel and Q. Y. J. Smithwick. Unique features of optical scanning, single fiber endoscopy. *Lasers in surgery and medicine*, 30(3):177–183, 2002.

[33] X. Li, D. J. MacDonald, and M. T. Myaing. Scanning fiber-optic nonlinear optical imaging and spectroscopy endoscope, January 17 2007. US Patent App. 11/623,974.

[34] M. T. Myaing, D. J. MacDonald, and X. Li. Fiber-optic scanning two-photon fluorescence endoscope. *Optics letters*, 31(8):1076–1078, 2006.

[35] P. Delaney and M. Harris. *Fiber-optics in scanning optical microscopy*. Springer, 2006.

[36] M. R. Harris et al. Scanning microscope with miniature head, November 22 2005. US Patent 6,967,772.

Bibliography

[37] W. Liang, K. Murari, Y. Zhang, Y. Chen, M.-J. Li, and X. Li. Increased illumination uniformity and reduced photodamage offered by the lissajous scanning in fiber-optic two-photon endomicroscopy. *Journal of Biomedical Optics*, 17(2):0211081–0211085, 2012.

[38] P. Xi, Y. Liu, and Q. Ren. Scanning and image reconstruction techniques in confocal laser scanning microscopy. *Laser Scanning, Theory and Applications, C.-C. Wang, ed.(Intech Open, 2009)*, pages 523–542.

[39] L. D. Swindle, S. G. Thomas, M. Freeman, and P. M. Delaney. View of normal human skin in vivo as observed using fluorescent fiber-optic confocal microscopic imaging. *Journal of investigative dermatology*, 121(4):706–712, 2003.

[40] L. E. Meyer and J. Lademann. Application of laser spectroscopic methods for in vivo diagnostics in dermatology. *Laser Physics Letters*, 4(10):754–760, 2007.

[41] J. Lademann, N. Otberg, H. Richter, L. Meyer, H. Audring, A. Teichmann, S. Thomas, A. Knüttel, and W. Sterry. Application of optical non-invasive methods in skin physiology: a comparison of laser scanning microscopy and optical coherent tomography with histological analysis. *Skin Research and Technology*, 13(2):119–132, 2007.

[42] V. Czaika, A. Alborova, W. Sterry, J. Lademann, and S. Koch. Application of laser scan microscopy in vivo for wound healing characterization. *Laser Physics Letters*, 7(9):685, 2010.

[43] A. Teichmann, U. Jacobi, E. Waibler, W. Sterry, and J. Lademann. An in vivo model to evaluate the efficacy of barrier creams on the level of skin penetration of chemicals. *Contact Dermatitis*, 54(1):5–13, 2006.

[44] A. Alborova. *Untersuchungen zur Wundheilung: Grundlagenuntersuchungen zum Wundheilungsverlauf mit und ohne Einwirkung von wassergefilterter Infrarot-A-Strahlung*. PhD thesis, Berlin, Charité, Univ.-Med., Diss., 2010, 2010.

[45] H. Zhai and H. I. Maibach. *Dermatotoxicology*. Informa Healthcare, 2004.

[46] S. Purwins, K. Herberger, E. S. Debus, S. J. Rustenbach, P. Pelzer, E. Rabe, E. Schaefer, R. Stadler, and M. Augustin. Cost-of-illness of chronic leg ulcers in germany. *International wound journal*, 7(2):97–102, 2010.

[47] J. Lademann, A. Kramer, L. Meyer, A. Alborova, W. Sterry, and B. Lange-Asschenfeldt. Characterization of wound healing by in vivo laser scanning microscopy. *GMS Krankenhaushygiene Interdisziplinär*, 4, 2009.

[48] G. E. Nilsson. Measurement of water exchange through skin. *Medical and Biological Engineering and Computing*, 15(3):209–218, 1977.

[49] H.-J. Weigmann, J. Ulrich, S. Schanzer, U. Jacobi, H. Schaefer, W. Sterry, and J. Lademann. Comparison of transepidermal water loss and spectroscopic absorbance to quantify changes of the stratum corneum after tape stripping. *Skin Pharmacology and Physiology*, 18(4):180–185, 2005.

[50] M. Breternitz, M. Flach, J. Prässler, P. Elsner, and J. W. Fluhr. Acute barrier disruption by adhesive tapes is influenced by pressure, time and anatomical location: integrity and cohesion assessed by sequential tape stripping; a randomized, controlled study. *British Journal of Dermatology*, 156(2):231–240, 2007.

[51] T. Vergou, S. Schanzer, H. Richter, R. Pels, G. Thiede, A. Patzelt, M. C. Meinke, W. Sterry, J. W. Fluhr, and J. Lademann. Comparison between tewl and laser scanning microscopy measurements for the in vivo characterization of the human epidermal barrier. *Journal of biophotonics*, 5(2):152–158, 2012.

[52] J. Krutz. *Qualitativer und quantitativer Vergleich zwischen dem kLSM Stratum und dem Vivascope 1500*. PhD thesis, Freie Universität Berlin, 2009.

[53] E. B. Podgorsak. *Review of radiation oncology physics a handbook for teachers and students*. 2003.

[54] BfArM (Bundesinstitut für Arzneimittel und Medizinprodukte). http://www.bfarm.de/en/bfarm/bfarm-node-en.html. 09/2012.

[55] R. Ding, E. Frei, M. Fardanesh, H. H. Schrenk, P. Kremer, and W. E. Haefeli. Pharmacokinetics of 5-aminofluorescein-albumin, a novel fluorescence marker of brain tumors during surgery. *The Journal of Clinical Pharmacology*, 51(5):672–678, 2011.

[56] J. T. Alander, I. Kaartinen, A. Laakso, T. Pätilä, T. Spillmann, V. V. Tuchin, M. Venermo, and P. Välisuo. A review of indocyanine green fluorescent imaging in surgery. *Journal of Biomedical Imaging*, 2012:7, 2012.

[57] J. Slavik. *Fluorescent probes in cellular and molecular biology*. CRC press Boca Raton, 1994.

[58] P. Kremer. *Albumin als Carrier zur laserinduzierten Fluoreszenzdiagnostik und Chemotherapie maligner Tumoren*. 2002.

[59] R. Ackroyd, C. Kelty, N. Brown, and M. Reed. The history of photodetection and photodynamic therapy. *Photochem Photobiol*, 74:656–669, 2001.

Bibliography

[60] A. Sharwani, W. Jerjes, V. Salih, A. J. MacRobert, M. El-Maaytah, H. S. M. Khalil, and C. Hopper. Fluorescence spectroscopy combined with 5-aminolevulinic acid-induced protoporphyrin ix fluorescence in detecting oral premalignancy. *Journal of Photochemistry and Photobiology B: Biology*, 83(1):27–33, 2006.

[61] R. Bonnett. Photosensitizers of the porphyrin and phthalocyanine series for photodynamic therapy. *Chemical Society Reviews*, 24(1):19–33, 1995.

[62] B. Lange-Asschenfeldt, A. Alborova, D. Krüger-Corcoran, A. Patzelt, H. Richter, W. Sterry, A. Kramer, E. Stockfleth, and J. Lademann. Effects of a topically applied wound ointment on epidermal wound healing studied by in vivo fluorescence laser scanning microscopy analysis. *Journal of Biomedical Optics*, 14(5):054001–054001, 2009.

[63] S. Astner, S. Dietterle, N. Otberg, H. J. Röwert-Huber, E. Stockfleth, and J. Lademann. Clinical applicability of in vivo fluorescence confocal microscopy for noninvasive diagnosis and therapeutic monitoring of nonmelanoma skin cancer. *Journal of biomedical optics*, 13(1):014003–014003, 2008.

[64] Thermo Fisher Scientific Inc. Fluorescence spectraviewer, last access 01/2014. *http://www.lifetechnologies.com/de/de/home/life-science/cell-analysis/labeling-chemistry/fluorescence-spectraviewer.html*.

[65] Richard P Haugland, Michelle TZ Spence, and Iain D Johnson. *Handbook of fluorescent probes and research chemicals*, volume 6. Molecular Probes Eugene, OR, 1996.

[66] M. T. Z. Spence and I. D. Johnson. *The Molecular Probes Handbook: A Guide to Fluorescent Probes and Labeling Technologies*. Live Technologies Corporation, 2010.

[67] J. R. Lakowicz. *Principles of Fluorescence Spectroscopy*, volume 3 of *Principles of Fluorescence Spectroscopy*. Springer, 2006.

[68] V. N. Mahajan. *Optical Imaging and Aberrations: Part 1. Ray Geometrical Optics*. 1998.

[69] H. Gross, H. Zügge, M. Peschka, and F. Blechinger. *Handbook of Optical Systems, Volume 3: Aberration Theory and Correction of Optical Systems*. Wiley, 2007.

[70] Laser Scan Lens Guide. Melles griot optics. *Rochester, NY*, 1987.

[71] J. W. Goodman. *Introduction to Fourier optics*. McGraw-Hill, New York, 1968.

[72] R. E. Hopkins. Optical system requirements for laser scanning systems. *Optics News*, 13(11):11–16, 1987.

[73] P. Belland and J. P. Crenn. Changes in the characteristics of a gaussian beam weakly diffracted by a circular aperture. *Applied Optics*, 21(3):522–527, 1982.

[74] H. Urey. Spot size, depth-of-focus, and diffraction ring intensity formulas for truncated gaussian beams. *Applied optics*, 43(3):620–625, 2004.

[75] L. Beiser. *Laser scanning systems*, volume 2. 1974.

[76] J. G. Skinner. Comment on light beam deflectors. *Applied Optics*, 3(12):1504–1504, 1964.

[77] M. A. Lauterbach. Finding, defining and breaking the diffraction barrier in microscopy–a historical perspective. *Optical Nanoscopy*, 1(1):1–8, 2012.

[78] J. Pawley. *Fundamental limits in confocal microscopy*. Springer, 2006.

[79] T. Wilson and A. R. Carlini. Size of the detector in confocal imaging systems. *Optics letters*, 12(4):227–229, 1987.

[80] M. Born and E. Wolf. *Principles of optics: electromagnetic theory of propagation, interference and diffraction of light*. Cambridge university press, 1999.

[81] P. J. Brosens. Dynamic mirror distortions in optical scanning. *Applied optics*, 11(12):2987–2989, 1972.

[82] T. Grasshoff. Finite element analysis of gimbal-mounted dual-axis mems mirror. (unpublished data), 2013.

[83] R. J. Noll. Zernike polynomials and atmospheric turbulence. *J. Opt. Soc. Am*, 66(3):207–211, 1976.

[84] H. Schenk, T. Sandner, C. Drabe, T. Klose, and H. Conrad. Single crystal silicon micro mirrors. *physica status solidi (c)*, 6(3):728–735, 2009.

[85] Y. Li. Performance analysis of laser projection systems of different configurations. *Optical Engineering*, 47(10):104301–104301, 2008.

[86] Y. Li. Single-mirror beam steering system: analysis and synthesis of high-order conic-section scan patterns. *Applied optics*, 47(3):386–398, 2008.

[87] M. Scholles, K. Frommhagen, C. Gerwig, J. Knobbe, H. Lakner, D. Schlebusch, M. Schwarzenberg, and U. Vogel. Recent advancements in system design for miniaturized mems-based laser projectors. 6911:69110U, 2008.

Bibliography

[88] Y.-Z. Zhang and D. Carter. Multicolor fluorescent microspheres as calibration standards for confocal laser scanning microscopy. *Applied Immunohistochemistry & Molecular Morphology*, 7(2):156–163, 1999.

[89] R. M. Zucker. Quality assessment of confocal microscopy slide based systems: performance. *Cytometry Part A*, 69(7):659–676, 2006.

[90] M. D. Abramoff, P. J. Magalhaes, and S. J. Ram. Image processing with imagej. *Biophotonics international*, 11(7):36–42, 2004.

[91] J. Hartmann. Objektivuntersuchungen. *Zeitschrift für Instrumentenkunde*, 24:1–21, 1904.

[92] B. C. Platt. History and principles of shack-hartmann wavefront sensing. *Journal of Refractive Surgery*, 17(5):S573–S577, 2001.

[93] R. V. Shack and B. C. Platt. Production and use of a lenticular hartmann screen. *J. Opt. Soc. Am*, 61(5):656, 1971.

[94] D. R. Neal, J. Copland, and D. A. Neal. *Shack-Hartmann wavefront sensor precision and accuracy*. 2002.

[95] G. Avwioro. Histochemical uses of haematoxylin - a review. *J Pharm Clin Sci (JPCS)*, 1:24–34, 2011.

[96] E. Gurr. *Synthetic dyes in biology, medicine and chemistry*. Elsevier, 1971.

[97] H. Gross, W. Singer, M. Totzeck, F. Blechinger, and B. Achtner. *Handbook of optical systems, Volume 4: Survey of Optical Instruments*. Wiley Online Library, 2005.

[98] H. G. J. Rutten, M. A. M. van Venrooij, R. Berry, and D. Lucas. *Telescope optics: A comprehensive manual for amateur astronomer*. Willmann-Bell, 1999.

[99] Schott AG. Tie-36: Fluorescence of optical glass. Technical report, Schott AG, 2005.

[100] Schott AG. Tie-35: Transmittance of optical glass. Technical report, Schott AG, 2005.

[101] C. von Ossietzky. Error theory and regression analysis. Technical report, Universität Oldenburg, Institute of Physics, 2013.

[102] H. H. Karow. *Fabrication methods for precision optics*. Wiley New York, 1993.

[103] P. R. Yoder. *Mounting optics in optical instruments*. 2008.

[104] D. Y. Wang, R. E. English Jr, and D. M. Aikens. Implementation of iso 10110 optics drawing standards for the national ignition facility. pages 502–508, 1999.

[105] J. Lane. Tutorial on iso 10110 optical drawing standard opti 521–intro to optomechanical engineering. 2009.

[106] J. Lane. A practical tutorial for generating iso 10110 drawings. 2012.